Lecture Notes in Computer S

Edited by G. Goos and J. Hartmanis

Advisory Board: W. Brauer   D. Gries   J. Stoer

561

Lecture Notes in Computer Science
Edited by G. Goos and J. Hartmanis

Advisory Board: W. Brauer D. Gries J. Stoer

C. Ding   G. Xiao   W. Shan

The Stability Theory of Stream Ciphers

Springer-Verlag

Berlin Heidelberg New York
London Paris Tokyo
Hong Kong Barcelona
Budapest

Series Editors

Gerhard Goos
Universität Karlsruhe
Postfach 69 80
Vincenz-Priessnitz-Straße 1
W-7500 Karlsruhe, FRG

Juris Hartmanis
Department of Computer Science
Cornell University
5148 Upson Hall
Ithaca, NY 14853, USA

Authors

Cunsheng Ding
Guozhen Xiao
Weijuan Shan
Dept. of Applied Mathematics & Inst. for Information Security
Xidian University, 710071, Xian, P. R. China

CR Subject Classification (1991): E.3, D.4.6, G.1.0

ISBN 3-540-54973-0 Springer-Verlag Berlin Heidelberg New York
ISBN 0-387-54973-0 Springer-Verlag New York Berlin Heidelberg

Typesetting: Camera ready by author
Printing and binding: Druckhaus Beltz, Hemsbach/Bergstr.
45/3140-543210 - Printed on acid-free paper

Preface

The problems of stabilizing control systems, linear and nonlinear systems, dynamical systems, adaptive systems and multivariable feedback systems as well as systems of differential equations are appealing and have been a motivating force for some years in the corresponding fields. The analysis of the stability of graphs and mappings as well as matrices also has engineering background. Some cryptosystems or cipher systems, as well as other kinds of systems, also have the problem of stabilization, though the cryptographic meanings of stability may vary for different cryptosystems.

This monograph reports our work in the field of the stability theory of stream ciphers which began in August 1987. To be self-contained, the research monograph also contains some known results with the purpose of employing them to support some new conclusions.

I would like to thank Prof. I. Ingemarsson, Dr. Xiangwu Liu and Prof. Xinmei Wang for their encouragement. I am indebted to Prof. T. Beth for his inspiration of my research interest in this field, to Prof. Tianshun Cao and Mrs Sulan Guo for their support of the monograph. I am particularly grateful to Prof. Yumin Wang who introduced me to information theory, and to the series reviewers for their commenting upon the manuscript and providing helpful suggestions. Many thanks are also due to the series editors and the executive editor for their interest in publishing the monograph, and to the copy-editor for correcting some mistakes in the manuscript.

It is impossible to sort out from all the references those that had a major influence on me; every such a partition is a distortion. Nevertheless, I am most aware of the influences of Prof. Guozhen Xiao, Prof.

T. Beth, Prof. J. L. Massey, Dr. R. A. Rueppel, Dr. T. Siegenthaler, Dr. D. Gollman, Prof. Yumin Wang and Ximei Wang.

Finally I wish to express my gratitute to the Chinese Natural Science Foundation and Xidian University for providing financial support
to the research project.

Should there be errors and bias in the monograph, it is I who am
responsible for all of them.

Xian, China   C. Ding
October 1991

Contents

1 Introduction

Since safeguarding communication and authenticating data have become more and more important, the need for cryptology research has become more necessary and urgent. Crypotology includes cryptography and cryptanalysis. The latter mainly deals with the breaking of cryptosystems. There are usually three kinds of cryptanalyses: *Ciphertext only attack*: only a piece of ciphertext is known to the cryptanalyst (and often the context of the message); *Known plaintext attack*: a piece of ciphertext with corresponding plaintext is known; *Chosen plaintext attack*: a cryptanalyst has a chosen piece of plaintext with corresponding ciphertext. For details, one can see the survey of recent results of cryptanalysis given by Brickell and Odlyzko [Brik 88]. Cryptography mainly deals with the investigation of methods for securing communications and authenticating data. As can be seen from the recent published literature, research on public-key cryptology and on stream ciphers as well as on authentication has been given the most consideration in recent years. A survey of recent research and development of public-key cryptology has been given by Diffe [Diff 88], and of authentication by Simmons [Simm 88], For details of the development of contemporary cryptology, one can consult the survey paper by Massey [Mass 88]. Since stream ciphers have historical and practical importance, they have been well investigated recently [Sieg 84, 85] [Ruep 86-88].

Concerning the general theory of data security, it is still not mature. At the workshop of Eurocrypt' 85, Prof. T. Beth called for results on lower bounds which would be the basis for an approach to a general theory of data security. He remarked [Beth 85]:

" The general appearance of many other 'Crypto-Schemes' and their immediate analysis shows, however, that we are still far away from a general theory. Even if we consider this problem optimistically, in my view it is clear that such a general theory would have to incorporate results on Complexity, Protocols and General Systems, which I count amongst the most difficult field of research at present. For research in complexity we urgently need results on lower bounds which would be the basis for an approach to a general theory of data security. The need for such a development has become especially obvious in the area of developing sequential ciphers. After the last few years successful work on designing PN-generators of large linear equivalent, it has now become apparent that other evaluation principle have to be applied".

Inspired by Dr. Beth's comments, the authors have tried to do research into some areas in correspondence with what Dr. Beth called for, including many bounds that may become parts of the basis for an approach to a general theory of data security, and to an evaluation principle (many measure indexes).

This research report is devoted to a new branch of stream ciphers: the stability theory of stream ciphers. It is mainly based on our research results, which have been obtained since 1987 and were mainly done by the first author. In order to be self-contained, the monograph also presents some known facts which will be useful in our analyses.

Chapter 2 gives an introduction to stream ciphers. Chapter 3 first introduces the two kinds of Walsh transforms and their properties. Then it discusses the best affine approximation of Boolean functions, which will be used as a basic tool for dealing with some problems of some of the following chapters. Finally, it presents the BAA attacks on two classes of stream ciphers.

Chapter 4 mainly introduces several measure indexes on the security of stream ciphers. Based on the results of Chapter 3, Section 4. 1 argues on whether correlation- immune functions are good filtering or combining functions for stream ciphers. Section 4.2 first shows some cryptographic merits and demerits of bent functions for some binary additive stream ciphers, then presents an autocorrelation characterization of bent functions. Section 4.3 introduces new measure indexes on the stability of linear complexity of sequences, i.e. , weight complexity or sphere surface complexity and sphere complexity, and also presents basic properties of the two measure indexes. Section 4.4 analyzes the security of several kinds of key-stream generators from the viewpoint of the best affine approximation attacks. Section 4.5 provides some results on the stability of elementary symmetric functions, since they are basic components of the GF(2)-interpretation of integer addition, which have been concluded to be useful in both public-key cryptosystems and stream ciphers.

Chapter 5 aims at investigating the stability of linear complexity of sequences. Section 5.1 provides basic results about the linear complexity of sequences. Section 5.2 is devoted to bounds on the weight complexities of binary sequences with period 2^n. Due to the importance of ML-sequences in stream ciphers, lower bounds on them are developed in Section 5.3. Based on the results of Section 5.3. , Section 5.4 cultivates lower bounds on the linear complexities of nonlinear- filtered ML-sequences. Since clock controlled ML-sequences have their merits as key streams, Section 5.5 develops bounds on the linear complexities of these sequences. Based on the merits of both clock controlled and non-linear-filtered binary ML-sequences, a new kind of key-stream generator is presented, and a lower bound on the linear complexity of the clock-controlled ML-sequences is derived. Because the linear-complexity stability of sequences is of great importance, Section 5.7 provides another approach to it by introducing another two measure indexes, i.e. , the fixed-complexity distance(FCD) and variable-complexity distance(VCD). Furthermore, the relationship between weight complexity and fixed-complexity distance as well as sphere complexity and variable-complexity distance are established by using Blahut's theorem.

Bounds on the fixed-complexity distance of binary sequences with period 2^n are also developed in this section.

Chapter 6 discusses the period stability of sequences, since the linear complexity stability of sequences has strong connections with their period stability. Section 6.1 provides general results about the order of polynomials and that of the period of sequences. Section 6.2 first gives, from the viewpoint of stream ciphers, two measure indexes on the period stability of sequences, i. e. , weight period and sphere period. Then it develops the relationships between weight period and weight complexity as well as sphere period and sphere complexity. Section 6.3 discusses some links between weight period and the auto-correlation functions of periodic sequences. Sections 6.4 and 6.5 are devoted to the development of some bounds on the weight period of some kinds of sequences. Chapter 7 first summarizes the monograph and presents nine open problems of the stability of stream ciphers, then introduces the concept and proposes some problems of the stability of source coding for the sources of binary additive stream ciphers.

We would like to make it clear that by the stability of stream ciphers, we take its narrow senses to mean the linear-complexity stability and period stability as well as the stability of their combining or filtering functions and their source codes. There may be some other indexes on the security or strength of stream ciphers, whose stabilities need to be investigated.

Since algorithms for computing the linear complexities of sequences are of importance to stream ciphers, two algorithms for fulfilling that task are provided in Appendices A and B. Appendix A presents Massey's conjectured algorithm for the LFSR synthesis of multi-sequences and gives a detailed proof. Furthermore, its applications to cryptology and coding are investigated. Appendix B presents a fast algorithm for the determination of linear complexities of sequences over GF (P^m) with period P^n, which is a generalization of the G-C algorithm.

2 Stream Ciphers

Cryptographic systems are generally classified into block and stream ciphers, in analogy to error-correcting codes which are subdivided into block and conventional codes. The clear distinction between block and stream ciphers is the memory (see Fig. 2. 1) [Mass 85, Ruep 86, Sieg 86].

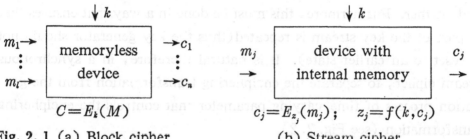

Fig. 2. 1 (a) Block cipher   (b) Stream cipher

A block cipher breaks plaintext message into successive blocks and enciphers each block M under control of a key k into a ciphertext block $C=(c_1, c_2, \cdots, c_n)$, where the message text alphabet and the ciphertext alphabet usually are identical. Each block is typically several characters long. Simple substitution and homophonic substitution ciphers are examples of block ciphers, even though the unit of encipherment is a single character. This is because the same key is used for each character. A stream cipher specifies a device with internal memory that enciphers the jth digit m_j of the message stream into the jth digit c_j of the ciphertext stream by means of a function which depends on both the secret key k and the internal state of the stream cipher at time j. A stream cipher is periodic if the key stream repeats after d characters for some fixed d; otherwise it is nonperiodic. Ciphers generated by Rotor and Hagelin machines are periodic stream ciphers. The Vernam cipher (one-time pad) is an example of nonperiodic stream ciphers.

There are two different approaches to stream encription: synchronous methods and self- synchronous methods. In synchronous stream ciphers, the next state depends only on the previous state and not on the input so that the succession of states is independent of the sequence of characters received, i. e. , the message stream. Consequently, the enciphering transformation is memoryless, but time-varying. But the device itself is not memoryless; it needs the internal memory to generate the necessary state sequence. This means that in synchronous stream ciphers, if a ciphertext character is lost during transmission, the sender and receiver must resynchronize their generators before they proceed further. Furthermore, this must be done in a way that ensures that no part of the key stream is repeated(thus the key generator should not be reset to an earlier state). It is natural therefore, in a synchronous stream cipher, to separate the enciphering transformation from the generation process of time-varying parameter that controls the enciphering transformation (see Fig. 2. 2).

The sequence $z^{\infty} = z_0 z_1 \cdots$ which controls the enciphering is called the key stream or running key. The deterministic automaton which produces the key stream from the actural key k and the internal state is called the running-key generator. Whenever the key k and the internal state are identical at the sender and receiver, the running keys necessarily are also identical and deciphering is easily accomplished.

In a self-synchronous stream cipher, each key character is derived from a fixed number n of the preceding ciphertext characters. Thus, if a ciphertext character is lost or altered during transmission, the error propagates forward for n characters, but the cipher synchronizes by itself after n correct ciphertext characters have been received (see Fig. 2. 3). Self-synchronous stream ciphers are nonperiodic because each key character is functionally dependent on the entire preceding message stream.

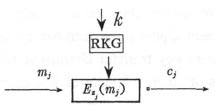

Fig. 2. 2 A Decomposed Synchronous Stream Cipher [Ruep 86]

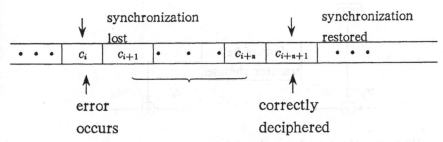

Fig. 2. 3 Propagation of error with self-synchronous stream ciphers

2. 1   Stream Cipher Systems Based on Exclusive-or Operation

One-time-pad (or binary Vernam cipher, see Fig. 2. 4) is the original model of modern stream cipher systems. The binary symmetric sources (BSS) are employed, in such a system, to produce uniformly distributed and statistically independent binary digits k_j, which are conveyed on a secure channel. The enciphering algorithm is the modulo-2 addition of plaintext digits m_j and key stream digits k_j, and the deciphering algorithm is the same. As pointed out by Shannon, this system is perfectly secure.

Actually, the one-time-pad cryptosystem is not practical, since the cost of transferring such a large number of key characters is enormous. What people expect are cryptosystems which have a finite number of keys and are convenient to use. But such systems are usually not perfectly secure, and are always only conditionally secure. Figure 2.5 represents such a stream cipher; the number of key characters to be transferred on the secure key transfer channel is relatively small. Historically, many kinds of stream ciphers of this class have been investigated.

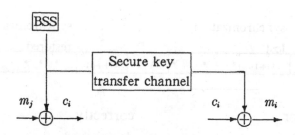

Fig. 2. 4   The one-time-pad cryptosystem

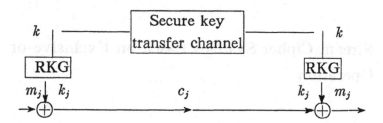

Fig. 2. 5   Stream cipher systems based on addition

2. 2   Finite State Machines and Key Stream Generators

Finite state machines constitute important mathematical objects for modelling electronic hardware specified above the register transfer level. Furthermore, due to their recursiveness finite state machines are convenient means for realizing infinite wordfunctions built over finite alphabets. It is well known that many functions of cryptographic systems can be modelled by finite state machines [Jenn 80, Pich 87, Ruep

86]. In a synchronous stream cipher, the running key generator may in general be viewed as autonomous finite state machines (see Fig. 2. 6).

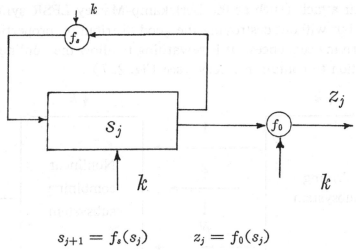

$$s_{j+1} = f_s(s_j) \qquad z_j = f_0(s_j)$$

Fig. 2. 6   The key stream generator as an autonomous finite state machine

The key stream generator as a finite state machine consists of an output alphabet and a state set, together with two functions and an initial state. The next state function f_s maps the current state S_j into a new state S_{j+1} from the state set, and the output function f_0 maps the current state S_j into an output symbol z_j from the output alphabet. The key may determine the next state function and the output function as well as the initial state. The fundamental problem of key stream generator design in the context of finite state machines is to find next state functions f_s and output functions f_0 which are guaranteed to produce a running key z^∞ that satisfies the basic requirements of large linear complexity, large period and uniform distribution properties. In order to fulfill these basic requirments, special classes of finite state machines have been employed as running key generators. Unfortunately, the theory of autonomous automata whose change of state function is nonlinear has not been well developed. When a linear autonomous automaton is combined with a nonlinear output mapping to be used as running key generator, Rueppel has subdivided the running key generator into a driving part and a combining part. The driving part then governs the state sequence of the running key generator and is responsible for providing sequences of large periods and good statistics. In contrast, the combining part controls the linear complexity of the key stream in order to make infeasible any linear attack (such as the Berlekamp-Massey LFSR synthesis algorithm), but without destroying the good distribution properties provided by the driving sequences. It is rewarding to allow the nonlinear combining function to contain memory (see Fig. 2.7).

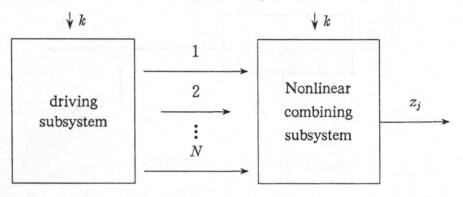

Fig. 2. 7 Running key generator subdivided as driving subsystem and nonlinear combining subsystem [Ruep 86]

There are some special kinds of important stream ciphers: the state-filtered generators, which employ one or several LFSRs, and the clock-controlled LFSR key stream generators, which use some LFSRs to control other LFSRs. For details, see Section 3. 3 and Section 5. 5.

2. 3 The Security of Stream Ciphers

The security of ciphers has two sides to its meaning, theoretical security and practical security. A cryptographic system is said to offer perfect secrecy (or to be unconditionally secure [Diff 79]), if the mutual information between the plaintext message and the associated ciphertext is zero, indepently of the length of the message. The one-time-pad is unconditionally secure, as proved by Shannon, since it is not difficult to see that the mutual information between the plaintext message and the associated ciphertext is zero. The one-time-pad is not practical because the cost of transferring the large number of key characters is very large. Consequently, steam ciphers with a finite key and convenient implementation have been developed, as modelled in Fig. 2. 5, which are conditionally secure.

The running key generator in Fig. 2. 2, controlled by the true key k, simulates a sequence which then is used to encipher the plaintext. The security of such a synchronous stream cipher now depends on the "randomness" of the key stream. In practice, the key stream is usually not truly random. Under known plaintext attacks, a cryptanalyst has a number of the running key characters. To make the system secure, the key stream must be unpredictable. There are now several requirements for unpredictability: a) *period requirement*: the key stream must have a long period, since the period defines a linear recursion; b) *linear complexity requirement*: it is necessary that the linear complexity(which is the length of the shortest linear feedback shift register able to produce the key) of the key stream be large enough, since there is an efficient LFSR synthesis algorithm for finding the shortest linear feedback shift register able to produce given sequences, provided that $2L$ consecutive digits of the sequence are given, where L is the linear complexity of the sequence; c) *independency requirement*: unpredictability requires that the next key stream digit "appears" to be drawn from a uniform distribution. Therefore, the key stream should have uniform statistics, i. e. , an equal distribution of single digits, of pairs, triples of digits, etc.

The foregoing three requirements for unpredictability are necessary, but badly insufficient. For example, let $s^\infty = (a_0 a_1 \cdots a_{2^m-3} \bar{a}_{2^m-2})^\infty$, where $(a_0 a_1 \cdots a_{2^m-2})^\infty$ is a maximum length binary sequence generated by a linear feedback shift register and $\bar{a}_{2^m-2} = 1 \oplus a_{2^m-2}$. The linear complexity of the sequence s^∞ is greater than or equal to $2^m - m - 1$, the period is $2^m - 1$, and s^∞ has a good distribution of 0-runs and 1-runs of various lengths. Nevertheless, if one knows $2m$

consecutive digits of s^∞, one can, with probability $1 - (2^m - 1)^{-1}$, construct a sequence which has probability $1 - (2^m - 1)^{-1}$ of agreement with s^∞. In other words, the linear complexity of the sequence is not " stable", i. e. , the linear complexity of the sequence decreases rapidly after changing only one digit bit in the corresponding place of every period segment of s^∞. In this research report we shall introduce several measure indexes for the unpredictability of sequences and the stability of stream ciphers generally.

3   The BAA Attacks on Several Classes of Stream Ciphers

Walsh functions and Walsh transforms have a wild range of applications to signal processing, image processing and communications as well as logic design and analysis [Beau 84, Mora 85]. The applications of spectral techniques to cryptology have been investigated by Xiao and Massey [Xiao 85] as well as Siegenthaler [Sieg 86] and Rueppel [Ruep 86]. In this chapter, two kinds of Walsh transforms and their basic properties are presented in Section 3. 1. Section 3. 2 presents the best affine approximation of Boolean functions, which will be used as a basic tool for dealing with some problems of some of the following chapters. Section 3. 3 presents the BAA attacks on two classes of stream ciphers which were developed by Ding and Xiao as well as Shan in October of 1987, and was presented at proceedings of the Third Chinese National Workshop on Cryptology in 1988 [Ding 87].

3. 1 Walsh Transforms and Their Properties

Since Walsh transforms will be used as basic tools for dealing with some problems of the stability of stream ciphers, we would like to introduce them and their basic properties, which will be useful in the following chapters. Due to the fact that there is a large amount of literature on the topic of Walsh-Fourier-Transforms, we shall present some theorems and facts about those transforms without proofs.

Let w and z be two vectors in $GF(2)^n$. For all w and z in $GF(2)^n$, the Walsh functions are defined as

$$Q(w,z) = (-1)^{wz} \tag{1}$$

where $wz = w_1z_1 \oplus \cdots \oplus w_nz_n$, and "$\oplus$" denotes modulo-2 addition (or exclusive-or operation). It is easy to see that all the Walsh functions are the characteristic functions of the Abelian group $(GF(2)^n,$ $\oplus)$.

It has been shown that any Boolean function $f : GF(2)^n \to GF(2)$ has a finite series expansion as follows [Karp 76]:

$$f(z) = \sum_{w \in GF(2)^n} S_f(w)Q(w,z) \tag{2}$$

with

$$S_f(w) = 2^{-n} \sum_{z \in GF(2)^n} f(z)Q(w,z) \tag{3}$$

Where $S_f(w)$ is called the first kind of spectrum of f , and the above transform is referred to as the first kind of Walsh transform. The second kind of Walsh transform is defined as

$$S_{(f)}(w) = 2^{-n} \sum_{z \in GF(2)^n} Q(w,z)(-1)^{f(z)} \tag{4}$$

with

$$f(z) = \frac{1}{2} - \frac{1}{2} \sum_{w \in GF(2)^n} S_{(f)}(w)Q(w,z) \tag{5}$$

It is easy to see that formula (5) holds. But from the definition of the first kind of spectrum it appears difficult to see what on earth the spectra of a Boolean function represent. The following Theorem 1 gives the relationship between the two kinds of spectrum of Boolean functions. By the following Theorem 3. 1 and the definition of the second kind of spectrum, we get the answer, i. e. , the spectra represent the extent to which a Boolean function correlates with all linear functions. **Theorem 3. 1** The above two kinds of spectrum are related by

$$S_{(f)}(w) = \begin{cases} -2S_f(w), & w \neq 0 \\ 1 - 2S_f(w), & w = 0 \end{cases}$$

Basic Properties:

$$(I) \sum_{w} S_{(f)}(w) = (-1)^{f(0)} \tag{6}$$

$$(II) \sum_{w} S_{f}(w) = f(0) \tag{7}$$

(III)*Parseval's theorem* [Tits 62]

$$\sum_{w \in GF(2)^{\bullet}} S_{f}(w)^{2} = 2^{n} \sum_{z \in GF(2)^{\bullet}} (f(z)^{2} \tag{8}$$

(IV)*Energy Conservation Law*[Mora 79]

$$\sum_{w \in GF(2)^{\bullet}} S_{(f)}(w)^{2} = 1 \tag{9}$$

(V) Let $F(w): GF(2)^{n} \to R$ be a function, where R is the real
number field. Then $F(w)$ is the second Walsh transform of
a binary Boolean function if and only if the following equations holds:

$$\sum_{w \in GF(2)^{\bullet}} F(w)F(w \oplus v) = \delta(v), \quad v \in GF(2)^{n} \tag{10}$$

where $\delta(v) = 1$ if $v = 0$; otherwise $\delta(v) = 0$ [Tits 62].

(VI)If the weight W_{f} of a Boolean function f is odd, then for all w
$\in GF(2)^{n}$, $S_{f}(w) \neq 0$. [Tits 62]

3. 2 The Best Affine Approximation of Boolean Functions

The best affine approximation (BAA) of Boolean functions has
wide applications in logic design [Mora 85]. By employing the BAA
approach, the realization complexity of circuits can be much reduced
without time penalty. It appears that its application to cryptology was
first noticed by Rueppel [Ruep 86]. He used the BAA to approximate

the S-box S_2 in the Data Encryption Standard (DES), but did not exploit its application to cryptology further. Since the best affine approximation of Boolean functions is the basic tool for our BAA approach to the cryptanalysis of some binary additive stream ciphers, in this section we present some facts about it.

Definition 3. 1 If the affine function $wx \oplus l$ makes the following formula achieve its minimal value,

$$\sum_{x \in GF(2)^n} (f(x) \oplus wx \oplus l), \qquad w \in GF(2)^n, \quad l \in GF(2)$$

then $wx \oplus l$ is called the best affine approximation of $f(x)$.

Theorem 3. 2 Let $P_f(wx \oplus l)$ denote the probability of agreement between $f(x)$ and $wx \oplus l$, then

$$P_f(wx) = \frac{1}{2} + \frac{1}{2} S_{(f)}(w), \qquad w \in GF(2)^n$$

and

$$P_f(wx) = \begin{cases} \dfrac{1}{2} - S_f(w), & w \neq 0 \\ 1 - S_f(w), & w = 0 \end{cases}$$

Proof: By the definition of $S_{(f)}(w)$, we have

$$\begin{aligned} S_{(f)}(w) &= 2^{-n} [\# \{x: f(x) = wx\} - \# \{x: f(x) \neq wx\}] \\ &= -1 + 2 \# \{x: f(x) = wx\} / 2^n \\ &= 2 P_f(wx) \end{aligned}$$

where $\# \{ \bullet \}$ denotes the number of elements in the set $\{ \bullet \}$. This proves the first part of Theorem 3. 2. The remaining part can be proved by using the result just proved and Theorem 3. 1.

The following Theorem 3. 3 about how to determine the best affine approximation of Boolean functions was developed by Rueppel with an information approach. It is also easy to see the result from the above Theorem 3. 2.

Theorem 3. 3 [Rueppel 86] Assume that

$$a = max\{ |S_{(f)}(w)| : w \in GF(2)^n\}$$

and $|S_{(f)}(w)| = a$. Then

i) If $S_{(f)}(w) \geqslant 0$, wx is a best affine approximation of $f(x)$ and the probability of agreement is

$$P_f(wx) = \frac{1}{2} + \frac{1}{2}a.$$

ii) If $S_{(f)}(w) < 0$, $1 \oplus wx$ is a best affine approximation of $f(x)$ and the probability of agreement is

$$P_f(1 \oplus wx) = \frac{1}{2} + \frac{1}{2}a.$$

Remark : The best affine approximation of a Boolean function is not unique. What we would like to find in logic design and cryptology is the one $wx \oplus l$ with minimum Hamming weight $W_H(w)$; the purpose of desiring such a linear function in logic design is for reducing the realization complexity of logic circuits. The cryptological reason is given in Section 4. 1. It also follows from Theorem 3. 2 that the spectra of $f(x)$ measure the correlation between $f(x)$ and all linear functions.

3. 3 The BAA Attacks on Two Classes of Stream Ciphers

The BAA approach to the analysis of the S-boxes of the Data Encryption Standard was given by Rueppel in 1986 [Ruep 86], but he did not exploit it further. In 1987 we applied this approach and some algebraic techniques together with some error-correcting techniques to the cryptanalysis of some stream ciphers [Ding 87]. In this section we shall introduce our analysis.

The following two classes of stream ciphers are of vital importance in the development of stream ciphers. Many stream ciphers are their deformations or originated from them. The concept of correlation immunity of the nonlinear combining function $f(x)$ was introduced to prevent the "divide and conquer" attack by Siegenthaler [Sieg 84]. Let x_1, x_2, $\cdots$, x_n be balanced i. i. d. (independent identically distributed) binary random variable, and $z = f(x_1, x_2, \cdots , x_n)$. $f(x)$ is said to be mth order correlation-immune if for each choice of indices i_1, i_2, $\cdots$,

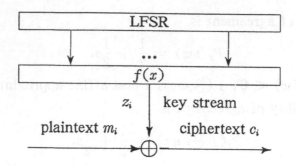

Fig. 3. 1 Nonlinear state filted stream ciphers

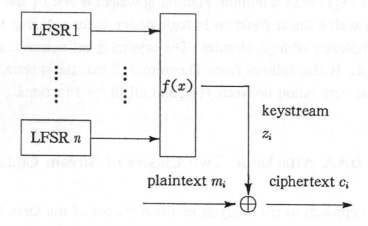

Fig. 3. 2 Nonlinear combined stream ciphers

i_m with $1 \leqslant i_1 < i_2 < \cdots < i_m \leqslant n$, the variable z is statistically independent of the vector $(x_{i_1}, x_{i_2}, \cdots, x_{i_m})$. This condition is, of course, equivalent to $I(x_{i_1}, x_{i_2}, \cdots, x_{i_m}; z) = 0$. In order to make the above two classes of stream ciphers secure, the linear complexity of the key stream must be large enough and the combining or filtering function $f(x)$ should possess certain correlation-immune order. Nevertheless, the fulfillment of the above two requirements does not guarentee a secure stream cipher. This will be shown clearly by the following analysis. Before giving the theoretical possibility of the BAA attack on the two classes of binary additive stream ciphers depicted in Fig. 3. 1 and Fig. 3. 2, we need the following lemmas:

Lemma 3. 4   Let $s_1^\infty, s_2^\infty, \cdots, s_t^\infty$ be t periodic binary sequences, then
$$L(s_1^\infty \oplus \cdots \oplus s_t^\infty) \leqslant L(s_1^\infty) + \cdots + L(s_t^\infty).$$
Especially, if $s_1^\infty, \cdots, s_t^\infty$ are t distinct phases of the same sequence s^∞, then
$$L(s_1^\infty \oplus \cdots s_t^\infty) = L(s_i^\infty) = L(s^\infty), \ 1 \leqslant i \leqslant t$$
where $L(s^\infty)$ denotes the linear complexity of s^∞.

Proof: It is known that for every periodic sequence s_i^∞, its generating function can be expressed as
$$s_i^\infty(x) = r_i(x)/f_i(x)$$
where $gcd(r_i(x), f_i(x)) = 1$ and $deg(r_i(x)) < deg(f_i(x))$ [see Chapter 5]. It follows from Theorem 5. 2 that $L(s_i^\infty) = deg(f_i(x))$. Therefore, we get

$$(\sum_{i=1}^{t} \oplus s_i^\infty)(x) = \sum_{i=1}^{t} \oplus s_i^\infty(x)$$

$$= \sum_{i=1}^{t} \oplus r_i(x)/f_i(x)$$

$$= \frac{\sum_{i=1}^{t} \oplus (\prod_{\substack{j \neq i \\ 1 \leqslant j \leqslant t}} f_j(x) r_i(x))}{\prod_{i=1}^{t} f_i(x)}$$

$$\triangleq g(x)/h(x)$$

Thus, we have

$$L(\sum_{i=1}^{t} \oplus s_i^\infty) = deg(h(x)) - deg(gcd(g(x), h(x)))$$

$$\leqslant deg(h(x))$$

$$= \sum_{i=1}^{t} deg(f_i(x))$$

$$= \sum_{i=1}^{t} L(s_i^\infty)$$

where $\sum_{i=1}^{t} \oplus s_i^\infty$ denotes the modulo-2 addition of s_1^∞, $s_2^\infty, \cdots, s_t^\infty$.

This proves the first part of Lemma 3. 4. The remaining part can be easily see from the above proof.

Lemma 3. 5   Let $s^n = s_1 \cdots s_n$ be a binary sequence, then the linear complexity of $\bar{s}^n = \bar{s}_1 \bar{s}_2 \cdots \bar{s}_n$ satisfies

$$L(s^n) - 1 \leqslant L(\bar{s}^n) \leqslant L(s^n) + 1$$

where $\bar{s}_i = 1 \oplus s_i$.

Proof : Assume that $(f(x), L)$ is the shortest LFSR that generates the sequence s^n , and

$$f(x) = 1 \oplus c_1 x \oplus \cdots \oplus c_L x^L$$

then by definition we have

$$a_n \oplus a_{n-1} c_1 \oplus \cdots \oplus a_{n-L} c_L = 0, \quad n \geqslant L + 1$$

Therefore we get the following two recursions

$$\bar{a}_n \oplus \bar{a}_{n-1} c_1 \oplus \cdots \oplus \bar{a}_{n-L} c_L = f(1)$$

Subtracting the above two equations we obtain

$$\bar{a}_{n+1} \oplus (1 \oplus c_1) \bar{a}_n \oplus (c_1 \oplus c_2) \bar{a}_{n-1} \oplus \cdots$$
$$\oplus (c_{L-1} \oplus c_L) \bar{a}_{n-L+1} \oplus \bar{a}_{n-L} c_L = 0$$

Hence, $L(\bar{s}^n) \leqslant L(s^n) + 1$. By symmetry we get $L(s^n) \leqslant L(\bar{s}^n) + 1$. This proves Lemma 3. 5.

Remark : From the above proof we see that $(f(x)(1 + x), L + 1)$ is a LFSR that generates $\bar{s}^n$ if $(f(x), L)$ is a shortest LFSR that generates s^n. We would also like to mention that the above lemma also holds for binary periodic sequences.

We now investigate the possibility of a BAA attack on the binary additive stream ciphers depicted in Fig. 3. 1. Suppose the length of the LFSR in Fig. 3. 1 is n , and the filter function $f(x)$ has m arguments. Let the BAA of $f(x)$ be $l(x) = x_{i_1} \oplus x_{i_2} \oplus \cdots \oplus x_{i_k} \oplus l$, where $k \leqslant n$, and $l \in GF(2)$, then the output sequence of the following sequence generator has probability of agreement with that of the original key stream generator $\frac{1}{2} + \frac{1}{2}a$, where $a = |S_{(f)}(w)|$ and w is a vector of $GF(2)^n$ such that $w_v = 1$ if v belongs to $\{i_1, \cdots, i_k\}$; otherwise $w_v = 0$. Because $S_{j_i}^\infty$, $i = 1, 2, \cdots, k$, are distinct phases of the output sequence of the original

LFSR, it follows from Lemma 3. 4 that the sequence generator in Fig. 3. 3 is equivalent to the following sequence generator of Fig. 3. 4 with feedback polynomial $m(x)$ or $m(x)(1+x)$, where $m(x)$ is the feedback polynomial of the original LFSR. Thus, if a is large enough, the output sequence of the generator in Fig. 3. 4 has high probability of agreement with that of the key stream generator. On the other hand, the sequence generator in Fig. 3. 4 is very simple to realize, since the length of the LFSR in Fig. 3. 4 is equal to or less than that of the original LFSR plus 1. The sequence generator in Fig. 3. 4 can be first used as a deciphering machine with the probability of correct decryption $\frac{1}{2} + \frac{1}{2}a$. Then one can correct the errors made by the deciphering machine by making use of the redundancy of language if the plaintext comes from a language source. The problem now is how to resume the sequence generator in Fig. 3. 4 by a BAA attack under the case of knowing a number of plaintext digits and the filter function.

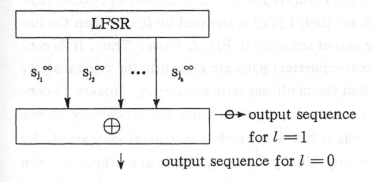

Fig. 3. 3   Sequence generator used to approximate the original generator

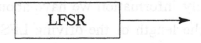

Fig. 3. 4   The reduced sequence generator of Fig. 3. 3

For the stream cipher dipicted in Fig. 3. 1, suppose the length of the LFSR is $n(n \geqslant L)$, choose $f(x_1, x_2, \cdots, x_L) = x_1 \oplus x_2 \oplus \cdots \oplus x_{L/2} \oplus x_{L/2+1} \cdots x_L$ as an example. Ruepple [Ruep 86] proved that if the filter function in Fig. 3. 1 has the form as given above, the linear complexity of the running key sequence is not less than $\binom{L}{L/2}$, which is approximately $(2\pi L)^{-\frac{1}{2}} 2^{L+1} (L \gg 1)$. Note the function $(2\pi L)^{-\frac{1}{2}} 2^{L+1}$ increases almost exponentially with L, and the linear complexity of the key stream in Fig. 3. 1 can be made as large as desired. On the other hand, $f(x)$ is balanced and is correlation-immune of order $\frac{1}{2}L - 1$. According to the three requirements (linear complexity requirement, correlation-immunity requirement and balance requirement) $f(x)$ is a good filter function. But by calculation we get

$$a = max|S_{(f)}(w)| = 1 - 2^{1-\frac{1}{2}L}$$

and the BAA of $f(x)$ is $x_1 \oplus x_2 \oplus \cdots \oplus x_{L/2} = l(x)$. The probability of agreement between $f(x)$ and $l(x)$ is $1 - 2^{-\frac{1}{2}L} > 99.8\% (L \gg 12)$. From Lemma 3. 4 we see that if $f(x)$ is replaced by $l(x)$, then the linear complexity of the output sequence in Fig. 3. 1 is n. Thus, if $2n$ consecutive bits of plaintext-ciphertext pairs are known to the cryptanalyst, he or she can expect that the nonlinear term $x_{L/2+1} \cdots x_L$ makes no contribution to the $2n$ bits of the key stream, since the probability of one contribution in the $2n$ bits is $2n2^{-\frac{1}{2}L}$, which is in general very small. By the well known Berlekamp-Massey LFSR synthesis algorithm, one can resume a LFSR, which can be used as a deciphering machine with the probability of correct decreption $1 - 2^{-\frac{1}{2}L}$.

We now show the above idea by a concrete example. For instance, suppose the only information we have about the kind of stream cipher in Fig. 3. 1 is the length of the driving LFSR, say 13, and the filter function $f(x) = x_1 \oplus x_2 \oplus \cdots \oplus x_6 \oplus x_7x_8 \cdots x_{12}$ as well as 26 consecutive bits of the key stream

$$s^{26} = 00000110111000111010111111$$

From the above analysis we can expect the nonlinear term $x_7 x_8 \cdots x_{12}$ makes no contribution to the 26 bits. By using the B-M algorithm, we obtain a new LFSR with feedback polynomial $x^{13} + x^4 + x^3 + x + 1$ which is the same as the original one.

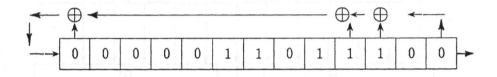

Fig. 3. 5 The constructed deciphering machine with correct decipher probability 98. 4%

Untill now we have completed the construction of a deciphering machine with a high probability of correct decryption. Furthermore, if we know that the ith variable x_i of the filter function takes the values of the j_ith storage cell of the original LFSR, i. e. , $f(x)$ taps at position $j_1, \cdots, j_m$, we can break the stream cipher completely. For the above example, suppose $j_i = i$ for $1 \leqslant i \leqslant 12$, let

$$V_0 = (a_0 a_1 \cdots a_{12})$$
$$V_1 = (a_1 a_2 \cdots a_{13})$$
$$\vdots \quad \vdots \quad \quad \vdots$$
$$V_{12} = (a_{12} a_{13} \cdots a_{24})$$

where V_0 is the state vector of the driving LFSR by which the filter function generates the first bit 0 of s^{26}, and $V_1, \cdots, V_{12}$ are the 12 consecutive state vectors that follows V_0. Since $a_n = a_{n-1} \oplus a_{n-3} \oplus a_{n-4} \oplus a_{n-13}$, every a_i can be expressed as a linear combination of $a_0, \cdots, a_{12}$. By solving the linear equations which contains 13 unknown variables a_0, a_1, $\cdots, a_{12}, \sum_{i=1}^{6} V_i = (0000011011100)$, we get $(a_0, a_1, \cdots, a_{12}) = (1100001100100)$. Since $V_0, V_1, \cdots, V_{12}$ are linearly independent, the solution of the above linear equations is unique. Hence the stream cipher is completely broken as depicted in Fig. 3. 6.

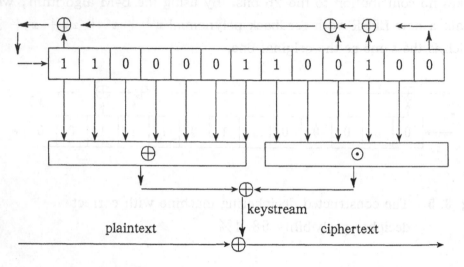

Fig. 3. 6   The completely broken stream cipher

That the foregoing BAA attack on the above stream cipher was successful is due to the fact that the nonlinear term $x_{L/2+1} \cdots x_L$ makes no contribution to the known 26 bits of key stream s^{26}. The probability with which the nonlinear term makes contributions M times to N consecutive bits of the key stream is equal to or less than $(N2^{-\frac{L}{2}})^M$. Although the probability is very small, it may occur. If some of the known N consecutive bits have contributions by nonlinear terms and the number of contributions M is small enough, one may elimilate those contributions by an error-correcting approach or the linear check approach given by Zeng [Zeng 87], under the condition that the feedback polynomial of the original LFSR is known. We now illustrate Zeng's method by an example given by Zeng.

Assume that $a^N = a_0 a_1 \cdots a_{N-1}$ is a finite binary sequence which satisfies the linear recursion

$$a_i \oplus a_{i+3} \oplus a_{i+20} = 0;$$

this means that $f(x) = 1 \oplus x^3 \oplus x^{20}$ is the feedback polynomial of a

LFSR that generates a^N. Let $b^N = b_0 b_1 \cdots b_{N-1}$ and $b^N = a^N \oplus e^N$, where $e^N = e_0 e_1 \cdots e_N$. Noticing that $f(x)^2 = 1 \oplus x^6 \oplus x^{40}$ and $f(x)^4 = 1 \oplus x^{12} \oplus x^{80}$ are also feedback polynomials of s^N, we see that the following 9 linear equations hold:

$$
\begin{aligned}
a_i &= a_{i+3} \oplus a_{i+20} \\
a_i &= a_{i-3} \oplus a_{i+17} \\
a_i &= a_{i-20} \oplus a_{i-17} \\
a_i &= a_{i+6} \oplus a_{i+40} \\
a_i &= a_{i-6} \oplus a_{i+34} \\
a_i &= a_{i-40} \oplus a_{i-34} \\
a_i &= a_{i+12} \oplus a_{i+80} \\
a_i &= a_{i-12} \oplus a_{i+68} \\
a_i &= a_{i-80} \oplus a_{i-68}
\end{aligned}
\tag{11}
$$

Assume we know b^N, but not a^N. One can expect to recover a number of bits of a^N from b^N if $W_H(e^N)/N$ is small enough. For fixed i, let

$$
\begin{aligned}
b_{i+3} \oplus b_{i+20} &= u_1 \\
b_{i-3} \oplus b_{i+17} &= u_2 \\
b_{i-20} \oplus b_{i-17} &= u_3 \\
b_{i+6} \oplus b_{i+40} &= u_4 \\
b_{i-6} \oplus b_{i+34} &= u_5 \\
b_{i-40} \oplus b_{i-34} &= u_6 \\
b_{i+12} \oplus b_{i+80} &= u_7 \\
b_{i-12} \oplus b_{i+68} &= u_8 \\
b_{i-80} \oplus b_{i-68} &= u_9
\end{aligned}
\tag{12}
$$

and $u = (u_1, \cdots, u_9)$. Zeng's method to recover a_i is based on the majority-logic decoding rule: let $a_i = 1$ if $W_H(u) \geqslant 5$; otherwise $a_i = 0$. The probability of correct recovery of a_i is greater than 78% [Zeng 87], which is a function of $W_H(e^N)/N$.

Similarly, we can attack the binary stream cipher depicted in Fig. 3.2. Assume that the length of LFSR i is L_i, and the combining function of the stream cipher, for example, is $f(x) = x_1 \oplus \cdots \oplus x_{\frac{n}{2}} \oplus x_{\frac{n}{2}+1} \cdots x_n$. If $2(L_1 + \cdots + L_{\frac{n}{2}+1})$ consecutive bits of plaintext-ciphertext pairs are known, and the nonlinear terms of $f(x)$ make no contribution to the corresponding $2(L_1 + \cdots + L_{\frac{n}{2}+1})$ bits of the key stream, we can also construct a deciphering machine with high probability of correct decryption. Generally speaking, the BAA attack on the two classes of stream ciphers depicted in Fig. 3. 1 and Fig. 3. 2 can be described as in Fig. 3. 7.

Comments and Remarks upon the BAA Attack

i) The BAA attack presented in this section is a known plaintext attack. It also assume that the combining or filtering function $f(x)$ of the above two kinds of binary additive stream cipher is known to a cryptanalyst. If $f(x)$ is not known, one cannot attack them in this way.

ii) The basic idea of the attack is not to recover the key or the original key stream generator (recovering the key or the original key stream generator may be very difficult), but to construct a new generator with an output sequence nearly the same as the original key stream, i. e. , with high probability of agreement with the original key stream, by making use of information about the key stream generator (speaking specifically, a number of plaintext-ciphertext pairs, and the function $f(x)$).

(iii) The attack presented in Fig. 3. 7 may be successful or not, depending on $max|S_{(f)}(w)|$ and the number of bits of the key stream a cryptanalyst get, as well as which segment of the key stream the cryptanalyst obtain. From the foregoing analysis, we see that the output sequence of the constructed deciphering machine has approximately the probability $\frac{1}{2} + \frac{1}{2}max|S_{(f)}(w)|$ of agreement with the original key stream. The attack must be a failure provided that $max|S_{(f)}(w)|$ is not large enough. On the other hand, assume that $l(x)$ is the BAA of $f(x)$, then $f(x)$ can be expressed as $f(x) = l(x) + g(x)$. If in the known segment of the key stream there are a number ($2n$ for the stream cipher of Fig. 3. 1, and $2 (L_1 + \cdots + L_n)$ for the stream cipher of Fig. 3. 2) of consecutive bits has no contributions by $g(x)$, and $max|S_{(f)}(w)|$ is large enough, then the attack can be applied. If not so, one has to know a relatively large number of bits of the key stream together with the feedback polynomials of the driving LFSRs in order to elimilate some of the contributions by $g(x)$.

28

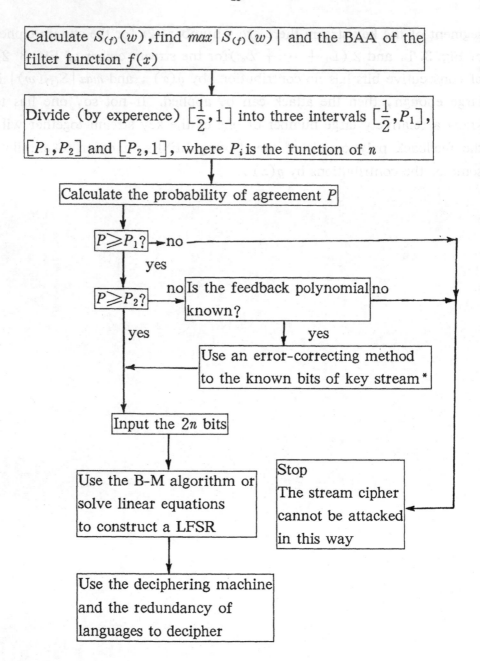

Calculate $S_{(f)}(w)$, find $max|S_{(f)}(w)|$ and the BAA of the filter function $f(x)$

Divide (by experence) $[\frac{1}{2},1]$ into three intervals $[\frac{1}{2},P_1]$, $[P_1,P_2]$ and $[P_2,1]$, where P_i is the function of n

Calculate the probability of agreement P

$P \geqslant P_1?$ → no

yes

$P \geqslant P_2?$ → no → Is the feedback polynomial known? → no

yes

yes

Use an error-correcting method to the known bits of key stream*

Input the $2n$ bits

Use the B–M algorithm or solve linear equations to construct a LFSR

Stop The stream cipher cannot be attacked in this way

Use the deciphering machine and the redundancy of languages to decipher

* : Such as Zeng's or another error-correcting approach.

Fig. 3. 7   A flow chart of the BAA attack on some binary additive stream ciphers

4 Measure Indexes on the Security of Stream Ciphers

So far several indexes on the security of stream ciphers have been proposed in order to guarantee the strength or security of stream ciphers. It seems according to the known literature that the linear complexity of the key stream and the correlation-immune order of the filtering or combining function are of great importance. But the proposed indexes cannot guarantee the security of stream ciphers as shown by Section 3. 3. Thus, other new indexes have to be introduced to measure the security of stream ciphers. This chapter is devoted to fufilling the task. Section 4. 1 argues on correlation-immune functions. Section 4. 2 shows the merit and demerit of bent functions as combining or filtering functions for some stream ciphers. Furthermore, some properties of bent functions are investigated in this section, including an autocorrelation characterization of bent functions and measure indexes on the stability of Boolean functions. Section 4. 3 introduces measure indexes on the stability of linear complexity of sequences and presents basic properties of the two measure indexes, i. e. , weight complexity or sphere surface complexity and sphere complexity. Section 4. 4 analyze the security of several kinds of key stream generator from the viewpoint of the best affine approximation attacks. As integer addition is useful in both public-key cryptosystems and stream cipher systems, and their GF(2)-interpretation is made up of elementary Boolean functions, Section 4. 5 gives some results on the stability of those functions.

4. 1   On Correlation-Immune Functions

Siegenthaler introduced the concept of correlation immunity of combining functions for nonlinear combined stream ciphers of Fig. 3. 2 (for definition, see Section 3. 3), and investigated the properties of Boolean functions with correlation immunity [Sieg 84, 86] . Xiao and Massey gave a spectral characterization of Boolean functions with correlation immunity [Xiao 85]. The structure and construction of correlation-immune functions (briefly, CI functions) were also studied by Shan [Shan 87]. The purpose of introducing correlation-immune functions as nonlinear functions for stream ciphers is to spare them from the "divide and conquer" attack [Sieg 85]. In this section we shall present our analysis about the suitability of correlation-immune functions for some stream ciphers [Ding 87]. Speaking specifically, this section aims at: (1) analyzing whether functions with correlation immunity are "good" combining or filtering functions for stream ciphers; (2) investigating the extent of gain and loss while employing CI functions as their combining or filtering functions; (3) showing by employing what functions as the combining functions how the binary additive stream cipher of Fig. 3. 1 can resist both the BAA and the "divide and conquer" attacks.

4. 1. 1   From the Energy-Conservation Law and the BAA Attack Viewpoints

To analyze CI functions from these viewpoints, we need the following theorems:

Lemma 4. 1 (Xiao-Massey)   A Booleam function $f: GF(2)^n \rightarrow GF(2)$ is correlation-immune of order m iff $S_f(w) = 0$ for all w with $1 \leqslant W_H(w) \leqslant m$, where $W_H(w)$ denotes the Hamming weight of w.

Theorem 4. 2   Let $f(x): GF(2)^n \rightarrow GF(2)$ be a Boolean function, then

there exists at least one affine function $wx \oplus l$ such that the probability
of agreement between $f(x)$ and $wx \oplus l$, denoted here and hereafter as
$P_f(wx \oplus l)$, is equal to or greater than $\frac{1}{2} + 2^{-\frac{n}{2}-1}$.

Proof : By the Energy Conservation Law (see Section 3.1):

$$\sum_{w \in GF(2)^n} S_{(f)}(w)^2 = 1$$

it follows that there exists at least one spectrum $S_{(f)}(w)$ of $f(x)$ such
that $|S_{(f)}(w)| \geqslant 2^{-\frac{n}{2}}$. Combining this result with Theorem 3.2, we
get

$$P_f(wx \oplus 1) \geqslant \frac{1}{2} + 2^{-\frac{n}{2}-1}$$

or

$$P_f(wx) \geqslant \frac{1}{2} + 2^{-\frac{n}{2}-1}$$

This proves Theorem 4.2.

Theorem 4.3 Let
$$P_f(wx, 1 \oplus wx) = max \{P_f(wx), P_f(1 \oplus wx)\}$$
then we have
$$\sum_{w \in GF(2)^n} (2P_f(wx, 1 \oplus wx) - 1)^2 = 1$$
Proof: By Theorem 3.3, we get
$$P_f(wx, 1 \oplus wx) = \frac{1}{2} + \frac{1}{2}|S_{(f)}(w)|$$
Thus, we obtain
$$|S_{(f)}(w)| = 2P_f(wx, 1 \oplus wx) - 1$$
Combining the Energy Conservation Law and the above result, we have
$$\sum_{w \in GF(2)^n} (2P_f(wx, 1 \oplus wx) - 1)^2 = 1$$

Theorem 4.4 If $f(x)$ is correlation-immune of order m, then there
exists at least one affine function $wx \oplus l$ such that

$$P_f(wx \oplus l) \geqslant \frac{1}{2} + \frac{1}{2}(2^n - \sum_{i=0}^{m} \binom{n}{i}))^{-\frac{1}{2}}$$

Proof : Combining Xiao-Massey's theorem and Theorem 3. 1, we see that a Boolean function $f(x)$ is correlation-immune of order m if and only if $S_{(f)}(w) = 0$ for all w with $1 \leqslant W_H(w) \leqslant m$. Therefore, it follows from the Energy Conservation Law that

$$\sum_{W_H(w)>m} S_{(f)}(w)^2 = 1$$

Thus, there must exist one $|S_{(f)}(w)|$ such that

$$|S_{(f)}(w)| \geqslant (\sum_{i=m+1}^{n} \binom{n}{i}))^{-\frac{1}{2}}$$

$$= (2^n - \sum_{i=0}^{m} \binom{n}{i}))^{-\frac{1}{2}}$$

By Theorem 3. 3, we have

$$P_f(wx \oplus l) \geqslant \frac{1}{2} + \frac{1}{2}(2^n - \sum_{i=0}^{m} \binom{n}{i}))^{-\frac{1}{2}}$$

for $l=0$ or 1.

From the Xiao-Massey theorem and Theorem 3. 1, we know that $f(x)$ is correlation-immune of order m if and only if $S_{(f)}(w) = 0$ for all w with $1 \leqslant W_H(w) \leqslant m$. But the second kind of spectrum of a Boolean function is constrained by not only the Energy Conservation Law, but also the following equations:

$$\sum_{w \in GF(2)^n} S_{(f)}(w)S_{(f)}(w \oplus v) = 0 , \qquad v \neq 0$$

Therefore, for a Boolean function $f(x)$ with correlation-immune order m, the probability of agreement between $f(x)$ and its BAA may be much higher than the lower bound given in Theorem 4. 4. On the other hand, Theorem 4. 3 means that the probabilities of agreement between a Boolean function and all affines functions are conservative. Thus, from the Energy Conservation Law and the BAA attack viewpoints, Boolean functions with certain correlation-immune orders may not be ideal combining or filtering functions for some stream ciphers. For instance, let us observe correlation-immune functions with arguments of 2, 3 and 4. Siegenthaler has given all the inherently distinct Boolean functions with correlation—immunity of arguments 2 and 3 as well as 4 [Sieg 86]. Since $max_w |S_{(f)}(w)| \geqslant \frac{1}{2}$ for all those functions, we conclude that they are nearly useless in stream ciphers.

4. 1. 2 From the Necessity Viewpoint

Before drawing conclusions about the suitability of CI functions for some stream ciphers, let us first recall how they were introduced. For the stream cipher depicted in Fig. 3. 2, assume that the length of $LFSR_i$ is r_i, $i=1,2,\cdots,n$, and the feedback polynomials of all LFSRs are primitive. Let R_i denote the number of different primitive feedback polynomials with degree r_i. Suppose the combining function $f(x)$ is known, then the total number of keys of the stream cipher is

$$K = \prod_{i=1}^{n} R_i(2^{r_i} - 1)$$

In a brute force attack and a worst case situation all of the keys have to be applied, which is by definition not feasible for a computationally secure PN-generator. By analyzing the correlation between $z=f(x)$ and the input variable x_i, Siegenthaler gave the "divide and conquer" attack which significantly reduces the number of trials from K to approximately $\sum_{i=1}^{n} R_i 2^{r_i}$. First, denoting $P_e^{(i)} = P(z=x_i), i=1,\cdots,n$, we see from Siegenthaler's analysis [Sieg 85] that the "divide and conquer" attack is not feasible if $P_e^{(i)} = \frac{1}{2}$ for every $1 \leqslant i \leqslant n$, which is equivalent to $f(x)$ being correlation-immune of order 1. But avoiding the "divide and conquer" attack does not necessarily require the combining function to be correlation-immune of order 1. The number of ciphertext digits needed to fulfill the attack can be made as large as one desires if

$|P_e^{(i)} - \frac{1}{2}|$ is small enough. Second, although the "divide and conquer" attack reduces the number of trials from K to $\sum\limits_{i=1}^{n} R_i 2^{r_i}$, this number may be large enough to guarantee a secure key stream generator provided that r_i, $1 \leqslant i \leqslant n$, are large enough. This shows that the "divide and conquer" attack may not really threaten the stream cipher. Thus, from the above analysis, we see that desiring the combining function to be correlation-immune may not be necessary. Consequently, we recommend Boolean functions such that

  i) $S_{(f)}(0) = 0$;

  ii) $|S_{(f)}(w)|$ are almost equal to each other for all $w \neq 0$;

  iii) their nonlinear order is large to some extent,

to stream ciphers as combining or filtering functions. Stream ciphers such as those depicted in Fig. 3. 1 and Fig. 3. 2 with combining or filtering functions of this class can resist both the BAA and the "divide and conquer" attacks. We prefer them to functions with correlation immunity, since the amount of mutual information between $z = f(x)$ and x_i are conservative and consequently information leakage cannot be avoided. Furthermore, it is well known that there is a trade-off between the nonlinear order and the correlation-immune order of Boolean functions.

We conclude from the above discussions that the following two indexes are reasonable in measuring the security of stream ciphers:

  a) $PV(f) = \max\limits_{w} |S_{(f)}(w)|$;

  b) $VS(f) = \sum\limits_{w \in GF(2)^n} (S_{(f)}(w)^2 - 2^{-n})^2$

$$= \sum\limits_{w \in GF(2)^n} S_{(f)}(w)^4 + (-1)^{f(0)} 2^{1-n} + 2^{-2n}$$

where we regard $\{S_{(f)}(w)^2, w \in GF(2)^n\}$ as a probability distribution of a random variable. The above two indexes can also be used as measure indexes for the stability of Boolean functions. We call a Boolean function $f: GF(2)^n \to GF(2)$ stable if $VS(f) = 0$ or equivalently

$PV(f) = 2^{-\frac{n}{2}}$, or $|S_{(f)}(w)| = 2^{-\frac{n}{2}}$ for every $w \in GF(2)^n$. The following result gives the relationship between the above two indexes.

Theorem 4. 5 Let $f(x)$ be a Boolean function of n arguments, then the two indexes defined above satisfy the following inequalities

$$(PV(f)^2 - 2^{-n})^2 \leqslant VS(f) \leqslant (2^{-\frac{1}{2}n}PV(f)^2 - 2^{-\frac{n}{2}})^2$$

Proof : It follows from the definition that

$$VS(f) = \sum_w (S_{(f)}(w)^2 - 2^{-n})^2$$
$$\leqslant 2^n(PV(f)^2 - 2^{-n})^2$$
$$= (2^{\frac{1}{2}n}PV(f)^2 - 2^{-\frac{1}{2}n})^2$$

The remaining part of Theorem 4. 5 is obvious.

4. 1. 3   From the Loss-and-Gain Viewpoint

Let us now analyze what is the gain and loss when we employ CI functions as combining or filtering ones for the stream ciphers of Fig. 3. 1 and Fig. 3. 2. First, let us observe the following function

$$f(x_1, x_2, x_3, x_4,) = x_3 + x_4 + x_1x_2 + x_1x_3 + x_1x_4$$
$$+ x_2x_3 + x_2x_4 + x_3x_4 + x_1x_2x_3$$
$$+ x_1x_2x_4 + x_1x_3x_4 + x_2x_3x_4$$

Its spectra are as follows

w	0	1	2	3	4	5	6	7	8	9	10	11	12	13	14	15
$S_{(f)}(w)$	$\frac{1}{4}$	0	0	$-\frac{3}{4}$	0	$\frac{1}{4}$	$\frac{1}{4}$	0	0	$\frac{1}{4}$	$\frac{1}{4}$	0	$\frac{1}{4}$	0	0	$\frac{1}{4}$

By Xiao-Massey therom, we know that $f(x)$ is correlation-immune of order 1. By reordering the ordered spectra of $f(x)$, we get another function with natually ordered spectra

$$S_{(g)}(x) = [\tfrac{1}{4}, 0, 0, \tfrac{1}{4}, 0, \tfrac{1}{4}, \tfrac{1}{4}, 0, 0, \tfrac{1}{4}, \tfrac{1}{4}, 0, \tfrac{3}{4}, 0, 0, \tfrac{1}{4}]$$

Again by calculation we get $g(x) = f(x_1, x_2, x_3, x_4) + x_1$. Noticing that $S_{(f)}(w) \neq 0$ for every w with $W_H(w) = 1$, we know that $g(x)$ is not correlation-immune.

Is there any difference between the above $f(x)$ and $g(x)$? The answer depends on how we consider them. From the viewpoint of the "divide and conquer" attack, they do have differences as shown in what follows. But from that of the BAA attack or of the stability of functions, we will see that they have no difference. Generally, many Boolean functions which are not correlation-immune can be changed into correlation-immune functions by performing a linear transform on the input variables and adding a linear function. Theoretically, we have the following conclusion.

Theorem 4.6 Let $f(x_1, x_2, \cdots, x_n)$ be a Boolean function, and
$$T_f \overset{\triangle}{=} \{w : S_{(f)}(w) = 0\} \neq \Phi$$
If there is a nonsingular $n \times n$ matrix A over $GF(2)$ and a vector b in $GF(2)^n$ such that
$$T_f(A,b) \overset{\triangle}{=} \{wA + b : w \in T_f\}$$
$$\supseteq \{z : 1 \leqslant W_H(z) \leqslant m, \, z \in GF(2)^n\}$$
then the function $g(x) = f(x(A^{-1})^t) + bA^{-1}x^t$ is correlation—immune of order m.

Proof: Let
$$a(w) = S_{(f)}(wA + b)$$
then
$$\sum_{w \in GF(2)^n} a(w)a(w + v) = \sum_{w \in GF(2)^n} S_{(f)}(wA + b)S_{(f)}(wA + vA + b)$$
$$= \sum_{z \in GF(2)^n} S_{(f)}(z)S_{(f)}(z + vA)$$
$$= \delta(vA) = \delta(v)$$
Thus, it follows from basic property (v) of Walsh transforms in Section 3.1 that $\{a(w) : w \in GF(2)^n\}$ is the spectra of a Boolean function $g(x)$. Hence it follows from Formula (5) of Section 3.1 that

$$g(x) = \frac{1}{2} - \frac{1}{2} \sum_{w \in GF(2)^*} S_{(g)}(w)(-1)^{wx^t}$$

$$= \frac{1}{2} - \frac{1}{2} \sum_{w \in GF(2)^*} S_{(f)}(wA + b)(-1)^{wx^t}$$

$$= \frac{1}{2} - \frac{1}{2} \sum_{w \in GF(2)^*} S_{(f)}(wA + b)$$

$$\times (-1)^{(wA+b)A^{-1^t}x^t}(-1)^{bA^{-1^t}x^t}$$

$$= \frac{1}{2} - \frac{1}{2}(1 - 2f(x(A^t)^{-1})) \cdot (-1)^{bA^{-1^t}x^t}$$

$$= f(x(A^{-1})^t) + bA^{-1}x^t$$

where x^t denotes the transpose vector of x.

Although CI functions were originally suggested for the binary additive stream cipher of Fig. 3. 2, we now analyze the gain and loss of employing CI functions for the stream cipher of Fig. 3. 1. Suppose that the filter function $f(x_1, \cdots, x_n)$ of the stream cipher is correlation-immune of order m, and the driving LFSR has length n and feedback polynomial

$$C(x) = 1 + c_1 x + \cdots + c_n x^n$$

with $c_n \neq 0$. We first show how to construct an equivalent key stream generator which is of the same type as the original one only with a different initial state vector and filter function $g(x)$, which is not correlation-immune. The procedure is as follows:

Step 1: Compute the spectra of $f(x)$, and find

$$E_f = \{w : S_{(f)}(w) \neq 0\}$$

Step 2: Let

$$C \overset{\Delta}{=} \begin{bmatrix} 0 & 0 & \cdots & 0 & c_n \\ 1 & 0 & \cdots & 0 & c_{n-1} \\ 0 & 1 & \cdots & 0 & c_{n-2} \\ \cdot & \cdot & \cdot & \cdot & \cdot \\ \cdot & \cdot & \cdot & \cdot & \cdot \\ \cdot & \cdot & \cdot & \cdot & \cdot \\ 0 & 0 & \cdots & 1 & c_1 \end{bmatrix}$$

determine the matrix set $M_C = \{D: D = f(C)$ and D is non-singular, $f(x) \in GF(2)[x]\}$.

Step 3: Choose a matrix D in M_c such that the row vectors of D^t contain as many vectors of E_f as possible and $|S_{(f)}(D_i)|$ is as large as possible, where D_i, $1 \leqslant i \leqslant n$, is a row vector of D^t.

Step 4: Let $g(x) = f(xD^{-1})$ be the new filter function and $V_1 = V_0D$ be the initial state vector of the LFSR of the new generator, where V_0 is the initial state vector of the original LFSR.

The above four steps summarize the procedure. By equivalence of two generators we mean they have the same output sequence. We now prove the equivalence of the new generator constructed according to the foregoing steps to the original one. The first thing we would like to mention is that M_C is not empty, since C belongs to M_C. Denote the ith state vector of the original LFSR as x^i, then $x^i = x^{i-1}C$. If $y^i = x^iD$ and $D \in M_C$, then $y^i = x^iD = x^{i-1}CD = x^{i-1}DC = y^{i-1}C$. This is because D is a polynomial matrix of C, so $CD = DC$. Therefore $\{y^i\}$ can also be regarded as state vectors of the same LFSR. Let z^i denote the ith state vector of the LFSR of the new generator, then $z^i = x^iD$. It follows that $g(z^i) = g(x^iD) = f(x^iDD^{-1}) = f(x^i)$. This proves the equivalence.

Let us now illustrate the foregoing procedure by the following example. In the key stream generator of Fig. 4. 1, the filter function $f(x) = x_1 + x_3 + x_4 + x_2x_5$ is correlation-immune of order 2.

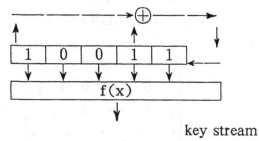

key stream

Fig. 4. 1　A nonlinear filtered key stream generator

First, by calculation we get $E_f = \{(11111), (11110), (10111), (10110)\}$ and $|S_{(f)}(w)| = \dfrac{1}{2}$ for every $w \in E_f$. Then we set

$$C = \begin{bmatrix} 0 & 0 & 0 & 0 & 1 \\ 1 & 0 & 0 & 0 & 0 \\ 0 & 1 & 0 & 0 & 0 \\ 0 & 0 & 1 & 0 & 1 \\ 0 & 0 & 0 & 1 & 0 \end{bmatrix}$$

and $D = I + C^3$, where I is the identity matrix. It is easy to check that D is invertable and

$$D^t = \begin{bmatrix} 0 & 1 & 0 & 0 & 1 \\ 1 & 0 & 1 & 1 & 0 \\ 0 & 1 & 0 & 1 & 1 \\ 1 & 0 & 1 & 1 & 1 \\ 1 & 1 & 0 & 1 & 1 \end{bmatrix}$$

Furthermore, the row vectors of D^t contain two vectors of E_f. Finally, we set $g(x) = f(xD^{-1}) = x_5 + x_2x_5 + x_3$ and $V_1 = (10011)D = (01001)$. The equivalent generator is as in Fig. 4. 2.

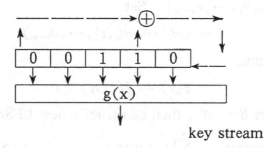

key stream

Fig. 4. 2 The equivalent key stream generator of Fig. 4. 1

Noticing that $S_{(g)}(00100) = S_{(g)}(00001) = \dfrac{1}{2}$, we see that $g(x)$ is not correlation-immune. Suppose that the keys of the stream ciphers in Fig. 3. 1 only consist of all the possible state vectors of its driving LFSR, the above example shows that stream ciphers with correlation-immune filter functions as depicted in Fig. 3. 1 are feasible to attack, provided that those with non-correlation-immune ones are feasible to attack. This means that employing correlation-immune filter functions in the stream cipher of Fig. 3. 1 may not get any advantage under the case the filter function and the feedback polynomial of the driving LFSR are known.

Stream ciphers of the type of Fig. 3. 2 with correlation-immune combining functions have also equivalent ones of the same type with non-correlation-immune combining functions. They can be found out by the following procedure:

Step 1: Calculate $\{S_{(f)}(w)\}$, and find the set

$$E_f = \{w : S_{(f)}(w) \neq 0, \quad w \neq 0\}$$

Step 2: Choose as many linearly independent vectors $\{w_1, \cdots, w_m\}$ in E_f as possible and also make $|S_{(f)}(w_i)|$ as large as possible, $1 \leqslant i \leqslant m$.

Step 3: Expand $\{w_1, \cdots, w_m\}$ as a basis for $GF(2)^n$, say, $\{w_1, \cdots, w_m, v_1, \cdots, v_{n-m}\}$. Set

$$A = (w_1^t, \cdots, w_m^t, v_1^t, \cdots, v_{n-m}^t)^t$$

and

$$g(x) = f(x(A^t)^{-1})$$

Step 4: Let $B = A^t$, then construct n new LFSRs such that

$$LFSRi' = \sum_{j=1}^{n} b_{ji} LFSRj, \qquad i = 1, 2, \cdots, n$$

After finishing the above four steps, an equivalent key stream generator is found, which is of the same type as the original one only with different driving LFSRs and combining function $g(x)$. Noticing that E_f is not empty, we see there is at least a vector w with $W_H(w) = 1$ and $S_{(g)}(w)$ $\neq 0$. Hence $g(x)$ is not correlation-immune. Since $f(x)$ is correlation-immune of order m, then $W_H(w) \geqslant m + 1$ for every $w \in E_f$. Some of the raw vectors of the nonsingular matrix thus have high Hamming weight, hence are the column vectors of matrix B. This generally led to the larger lengths of the newly constructed LFSRs compared with those of the original ones. This shows that employing combining functions with correlation immunity does have some advantage, but the advantage may be limited. As shown in Section 3. 3, the loss is that employing CI functions may cause the stream ciphers be open to a BAA attack.

(Note: The initial work for this section was done in 1987 [Ding 87] and presented at proceedings of the Third Chinese National Workshop on Cryptology in 1988.)

4. 2   The Cryptographic Merits and Demerits of Bent Functions

Rothaus first defined the concept of bent functions and discussed some of their properties [Roth 76]. The application of bent functions to signal protection was investigated by Olsen, Scholtz and Welch [Olsen 82], and Kumar as well as Scholtz [Kuma 83]. They presented a class of bent sequences with linear complexity greater than $\binom{n/2}{n/4} 2^{n/4}$, where n is the length of the shift register generating the m-sequence. In 1988 when we were looking for stable functions for stream ciphers, our college Dr. Cuankuen Wu noticed that bent functions are stable and pointed out their application to stream ciphers [Wu 88]. In this section we shall show the merit and demerit of employing bent functions as combining or filtering functions for the binary additive stream cipher of Figs. 3. 1 and 3. 2, then present some cryptographically useful properties of bent functions.

A Boolean function $f(x_1, x_2, \cdots, x_n) : GF(2)^n \rightarrow GF(2)$ is called bent if all the second spectra of $f(x_1, \cdots, x_n)$ are $\pm 2^{-\frac{n}{2}}$ [Roth 76]. Rothaus also constructed the following class of bent functions

$$f(x_1, \cdots, x_n, y_1, \cdots, y_n) = \sum_{i=1}^{n} x_i y_i + g(y_1, \cdots, y_n)$$

where $g(y_1, \cdots, y_n)$ is another Boolean function. Let the binary additive stream cipher of Fig. 3. 2 employing the above bent function as its combining one, i. e. , the $2n$ maximum-length LFSRs be combined by a bent function as shown above. If all the $2n$ LFSR lengths $L_1, L_2, \cdots, L_{2n}$ are different than two, then the key stream is guaranteed to have maximum linear complexity which is equal to $f(L_1, L_2, \cdots, L_{2n})$ provided that the $2n$ LFSRs are started in a nonzero state [Ruep 87]. Since the function $g(y_1, y_2, \cdots, y_n)$ is freely chosen, achieving relatively large linear complexity for the key stream is not difficult. Nevertheless, it has been shown that $deg(f) \leqslant n/2$ if $f(x_1, \cdots, x_n)$ is bent and $n > 2$. Therefore, the linear complexity of the key stream is limited to some extent.

For the binary additive stream cipher of Fig. 3. 1 employing a bent function as shown above, Kumar and Scholtz proved that the linear complexity of the key stream is lower bounded by $\binom{n/2}{n/4} 2^{n/4}$, where n is a multiple of 4. Thus, achieving large linear complexity for the key stream is feasible.

Since bent functions are stable, as shown in Sections 3. 3 and 4.

1, stream ciphers of Figs. 3.1 and 3.2 which employ bent functions can resist the BAA attack. On the other hand, since $P^{(i)} = P(z = x_i) = \frac{1}{2} \pm (\frac{1}{2})^{\frac{n}{2}+1}$, the "divide and conquer" attack on the stream cipher of Fig. 3.2 is also infeasible. Thus, on the whole, some bent functions are ideal combining and filtering functions for the stream ciphers of Figs. 3.1 and 3.2. However, another minor demerit of bent functions for stream ciphers is that they are not balanced, since $S_{(f)}(0) = 2^{-\frac{n}{2}}$. This can be overcome by modifying bent functions slightly as $g(x_1, \cdots, x_{n+1}) = f(x_1, \cdots, x_n) + x_{n+1}$, where $f(x_1, \cdots, x_n)$ is a bent function. Apparently, $g(x_1, \cdots, x_{n+1})$ is balanced, and $PV(g) = \max_{w} |S_{(g)}(w)| = 2^{-n/2-1}$. Thus, the modified bent functions are also ideal combining and filtering functions.

Since bent functions are useful in stream ciphers, we now present an autocorrelation characterization of bent functions. The autocorrelation function of a Boolean function is defined as

$$C_f(w) = 2^{-n} \sum_{x \in GF(2)^n} (-1)^{f(x)+f(x+w)}, \qquad w \in GF(2)^n$$

It follows from the above definition that $|C_f(w)|$ is the measure of the extent to which $f(x)$ correlates with $f(x+w)$.

Theorem 4.7 Let $f : GF(2)^n \rightarrow GF(2)$ be a Boolean function, then f is bent iff f satisfies the following 2^n equations

$$C_f(w) - \begin{cases} 1, & w = 0 \\ 0, & w \neq 0 \end{cases} \tag{1}$$

Proof: Assume that f is bent, then $S_{(f)}(w) = \pm 2^{-\frac{n}{2}}$. Let

$$S_{(f)}(w) = (-1)^{g(x)} 2^{-\frac{n}{2}}$$

then $g(x)$ is also a Boolean function. By formula (5) of Section 3.1, we get

$$f(x) = \frac{1}{2} - \frac{1}{2} \sum_{w \in GF(2)^n} S_{(f)}(w)(-1)^{wx}$$

$$= \frac{1}{2}(1 - 2^{-\frac{n}{2}} \sum_{w} (-1)^{wx+g(w)})$$

$$= \frac{1}{2}(1 - 2^{\frac{n}{2}} S_{(g)}(x))$$

whence, we have

$$S_{(g)}(x) = 2^{-\frac{n}{2}}(-1)^{f(x)}$$

Since $S_{(g)}(x)$ is the second kind of Walsh transform of $g(x)$, it follows from the basic property (v) of Section 3. 1 that $C_f(w) = 1$ if $w=0$; otherwise $C_f(w) = 0$.

Conversely, if f satisfies equation (1), then $2^{-\frac{n}{2}}(-1)^{f(x)}$ is the second kind Walsh spectrum of a Boolean function by the basic property (v) in Section 3. 1, say $g(x)$, i. e. , $S_{(g)}(x) = 2^{-\frac{n}{2}}(-1)^{f(x)}$. Similarly, we can prove that $S_{(f)}(x) = 2^{-\frac{n}{2}}(-1)^{g(x)}$. Hence $f(x)$ is a bent function.

The above Theorem 4. 7 is very useful in analyzing bent functions. First, it shows that the probability of agreement between $f(x)$ and $f(x+w)$ is $\frac{1}{2}$ for each $w \neq 0$. This may be another good property which is of cryptographic significance. Second, the theorem may be used as another way to enumerate bent functions, to construct bent functions and to investigate properties of bent functions. For instance, let us now employ it to prove the following result:

Theorem 4. 7a Let

$$f(x_1, \cdots, x_n) = \sum_{1 \leqslant i < j \leqslant n} a_{ij} x_i x_j + \sum_{i=1}^{n} b_i x_i$$

then $f(x)$ is bent iff the following $n \times n$ matrix A is nonsingular, where A is defined as

$$A = \begin{bmatrix} 0 & a_{12} & a_{13} & a_{14} & \cdots & a_{1,n-1} & a_{1n} \\ a_{12} & 0 & a_{23} & a_{24} & \cdots & a_{2,n-1} & a_{2n} \\ \cdot & \cdot & \cdot & \cdot & \cdot & \cdot & \cdot \\ \cdot & \cdot & \cdot & \cdot & \cdot & \cdot & \cdot \\ \cdot & \cdot & \cdot & \cdot & \cdot & \cdot & \cdot \\ a_{1,n-1} & a_{2,n-1} & a_{3,n-1} & a_{4,n-1} & \cdots & 0 & a_{n-1,n} \\ a_{1n} & a_{2n} & a_{3n} & a_{4n} & \cdots & a_{n-1,n} & 0 \end{bmatrix}$$

Proof: Since it is known that $f(x) + l(x)$ is bent if $f(x)$ is so and l (x) is linear, we only need to consider the Boolean function $h(x)$

$$h(x) = \sum_{1 \leqslant i < j \leqslant n} a_{ij} x_i y_j$$

Set

$$g_w(x) = h(x) + h(x + w)$$

then we obtain

$$g_w(x) = \sum_{i<j} (a_{ij} w_i x_j + a_{ij} w_j x_i) + a$$

$$= \sum_{j=2}^{n} \left(\sum_{i=1}^{j-1} a_{ij} w_i \right) x_j + \sum_{i=1}^{n-1} \left(\sum_{j=i+1}^{n} a_{ij} w_j \right) x_i + a$$

$$= \left(\sum_{j=1}^{n-1} a_{1i} w_i \right) x_1 + \left(\sum_{i=1}^{n-1} a_{in} w_i \right) x_n$$

$$\left| \sum_{j=1}^{n-1} \left(\sum_{i=1}^{j-1} a_{ij} w_i \right| \sum_{i=j+1}^{n} a_{ji} w_i \right) x_j \right| a$$

$$= x A w^t + a$$

where $a = \sum_{1 \leqslant i < j \leqslant n} a_{ij} w_i w_j$, $w^t = (w_1, \cdots, w_n)^t$ and $x = (x_1, \cdots, x_n)$.

It follows that $g_w(x) \neq a$ for each $w \neq 0$, which is equivalent to $Aw^t \neq$

0 for each $w \neq 0$. On the other hand, apparently $g_0(x) = 0$. Thus, by Theorem 4.7, $h(x)$ is bent iff A is nonsingular. This proves the theorem.

It was known that a bent function is simply a difference set in an elementary Abelian 2-group [Olse 82] [McFa 73]. To show the importance of Theorem 4.7, we employ it to present the relationship between bent functions and difference sets.

Theorem 4.7b Let $f(x_1, \cdots, x_n)$ be a Boolean function, and $f(x)^{-1} = \{c_1, \cdots, c_k\} = C$. Denote $C + a = \{c_1 + a, \cdots, c_k + a\}$. where a, $c_i \in GF(2)^n$. Then

(i) if $W_H(f) = k = 2^{n-1} + 2^{\frac{n}{2}-1}$, then $f(x)$ is bent

iff $\#(C \cap C_a) = 2^{n-2} + 2^{\frac{n}{2}-1}$ for each $a \neq 0$.

(ii) if $W_H(f) = k = 2^{n-1} - 2^{\frac{n}{2}-1}$, then $f(x)$ is bent

iff $\#(C \cap C_a) = 2^{n-2} - 2^{\frac{n}{2}-1}$ for each $a \neq 0$.

where $\#\{\cdot\}$ denotes the number of elements in the set $\{\cdot\}$, and $W_H(f)$ denotes the weight of $f(x)$.

Proof: Let $g_a(x) = f(x) + f(x + a)$. Noticing that $f(x)^{-1}(1) = C$ and $f(x)^{-1}(0) = GF(2)^n - C$ as well as $f(x + a)^{-1}(1) = C_a$, $f(x + a)^{-1}(0) = GF(2)^n - C_a$ we get

$$\#(g_a(x)^{-1}(1)) = \#((C \cap (GF(2)^n - C_a))$$
$$\cup (C_a \cap (GF(2)^n - C)))$$
$$= 2(W_H(f) - \#(C_a \cap C))$$

It follows from Theorem 4.7 that $f(x)$ is bent iff $g_a(x)$ is balanced for each $a \neq 0$, which is equivalent to $\#(g_a(x)^{-1}(1)) = 2^{n-1}$ for each $a \neq 0$. Hence $f(x)$ is bent iff $2(W_H(f) - \#(C_a \cap C)) = 2^{n-1}$, i.e., $\#(C \cap C_a) = W_H(f) - 2^{n-2}$. On the other hand, it is known that if f is bent, then n must be even and $W_H(f) = 2^{n-1} \pm 2^{\frac{n}{2}-1}$. Therefore, Theorem 4.7b holds.

Example 1 Let $n=6$ and $f(x)^{-1}(1) = \{c: c \in GF(2)^6$ and $W_H(c) = 0,1,4,5,\}$, then $W_H(f) = \binom{6}{0} + \binom{6}{1} + \binom{6}{4} + \binom{6}{5} = 28 = 2^{n-1} - 2^{\frac{n}{2}-1}$. It can be checked that for each $a \neq 0$, $\#(C \cap C_a) = 12$. Thus, $f(x)$ is a bent function.

Example 2 Let $n=10$ and $f(x)^{-1}(1) = \{c: c \in GF(2)^{10}$ and $W_H(c) = 2,3,6,7$ and $10\}$, then

$$W_H(f) = \binom{10}{2} + \binom{10}{3} + \binom{10}{6} + \binom{10}{7} + \binom{10}{10} = 2^{n-1} - 2^{\frac{n}{2}-1}$$

It can also be checked that for each $a \neq 0$, $\#(C \cap C_a) = 240$. Hence, $f(x)$ is a bent function.

Let $f(x): GF(2)^n \to GF(2)$ be a Boolean function, then $C = f(x)^{-1}(1)$ is called the characteristic set of the function f. Conversely, $f(x)$ is called the characteristic function of the set C. Thus there is a one-to-one correspondence between the subsets of $GF(2)^n$ and all binary Boolean functions of n arguments. The characteristic sets of bent functions are closely related with the so-called difference sets of $GF(2)^n$. Let $(G,+)$ be an Abelian group of order N, D a subset of G with k elements. Set $D^- = \{d_i - d_j : d_i, d_j \in D, d_i \neq d_j\}$. If each nonzero element of G appears exactly λ times in D^-, then D is called a (N,k,λ)-difference set. It is easy to see that a subset D of G is a (N,k,λ)-difference set iff $\#(D \cap D_a) = \lambda$ for each $a \neq 0$. Thus, it follows from Theorem 4.7b that the following result holds.

Theorem 4.7c Let $f(x): GF(2)^n \to GF(2)$ be a Boolean function. Then $f(x)$ is bent iff the characteristic set of $f(x)$ is a $(2^n, 2^{n-1} \pm 2^{\frac{n}{2}-1}, 2^{n-2} \pm 2^{\frac{n}{2}-1})$ -difference set.

4. 3 Weight Complexity (Sphere Surface Complexity) and Sphere Complexity

Stream ciphers employ random sequences, referred to as key streams, to encipher message streams. In order to make a stream cipher secure enough, the key stream must be as "random" as possible. Since the running key generator is usually a finite state machine, the key stream is necessarily (ultimately) periodic. Thus, we are facing the problem of measuring the randomness of both finite and infinite sequences. It is difficult to define adequately the concept of randomness mathematically for both finite and infinite sequences. D. E. Knuth [Knut 81] discusses various concepts of randomness for infinite sequences and gives a short description of how randomness of a finite sequence could be defined. Golomb proposed [Golo 67] the following three postulates to measure the randomness of a periodic binary sequence. First, the number of zeros and the number of ones are as equal as possible per period, i. e. , $p/2$ if p is even and $(p\pm1)/2$ if p is odd, where p is the period of a sequence. Second, a fraction $\frac{1}{2^i}$ of the total number of runs has length i, as long as there are at least 2 runs of length i. Third, the out-of-phase autocorrelation $AC(k)$ has the same value for all k. The first postulate states that the zeros and ones occur with roughly the same probability. The second implies that after 01 the symbol 0 has about the same probability as the symbol 1, etc. So it says that certain n-grams occur with the right probabilities. The interpretation of the third postulate is more difficult. It does say that counting the number of agreements between a sequence and a shift version of that sequence does not give any information about the period of that sequence, unless one shifts over a multiple of the period. Every sequence which satisfies these three random postulates is called a pseudo-noise (PN) sequence. Golomb's three random postulates do not define a general measure of randomness for sequences, since the maximum-length sequences or m-sequences satisfy all Golomb's three randomness postulates, but they are highly predictable. Let L denote the degree of the primitive minimal polynomial of a PN-sequence, then only $2L$ bits of the sequence suffice to specify completely the remainder of the period of length $2^L - 1$. It is natural that the idea of randomness should also reflect the impossibility of predicting the next digit of a sequence from all the previous ones. The finite state mchine approach to a definition of randomness of a finite sequence based on this concept of unpredictability was taken by R. Solomonov [Solo 64] and A. Kolmogorov [Kolm 65]. They characterized the " patternlessness" of a finite sequence by the length of the shortest Turing machine program that could generate the sequence. Patternlessness may be equated with unpredictability or randomness. Investigating the randomness or complexity of a sequence by itself instead of by the algorithm (the finite state machine approach), A. Lempel and J. Ziv [Lemp 76] defined a kind of complexity for sequences, which was referred to as apparent complexity by Lempel and Ziv. The linear complexity, the shortest linear feedback shift register (LFSR), was introduced to measure the linear unpredictability of a sequence (finite or periodic) and was regarded as an important index for the security of the key stream. This is because there exists an efficient LFSR synthesis algorithm, the Berlekamp-Massey algorithm [Mass 69].

Although linenar complexity is a necessary index for measuring the unpredictability of a sequence, clearly it is not sufficient. High linear complexity does not guarantee high unpredictability. In Sections 3. 3

and 3. 4, we have seen that although the linear complexity of a sequence may be very large, we can construct a new one with lower linear complexity such that the probability of agreement is large. The "linear complexity stability" of this kind of sequences is very bad, i. e. , after changing a few bits of the original sequence, its linear complexity decreases or increases fast. This also shows that sequences with high linear unpredictability may be well approximated by sequences with very lower linear unpredictability. The several classes of stream ciphers presented in Sections 3. 3 and 3. 4 were successfully attacked by the BAA approach under certain conditions becasuse the linear complexity of their key stream is not stable. In order to measure the stability of linear complexity and the unpredictability of sequences, we introduce the following indexes [Ding 87].

Definition 4. 8   Let x be a sequence of length n, then the weight complexity of x is defined as

$$WC_u(x) = \min_{W_H(y)=u} L(x+y)$$

where $W_H(y)$ denotes the Hamming weight of y, i. e. , the number of components of y that are different from zero. $L(x)$ denotes the linear complexity of the sequence x.

We now give a geometrical descreption of the weight complexity of binary sequences. Consider the space $GF(2)^n$, take the Hamming distance d_H as its distance. Denote $S(x,u)=\{y:d_H(x,y)=u\}$. By definition we know that

$$WC_u(x) = \min_{y \in S(x,u)} L(y)$$

This means that the weight complexity $WC_u(x)$ is the maximum lower bound of linear complexities of all the sequences with length n on the surface of the sphere $S(x,u)$. That is why we call it sphere surface complexity.

Definition 4. 9　Let $O(x,u)=\{y: 0<d_H(x,y)\leqslant u\}$ be the sphere without the center x. Define

$$SC_u(x) = \min_{y\in O(x,u)} L(y)$$

We call this the u-sphere complexity of the finite squence x.

The geometrical meaning of the u-sphere complexity is clear. It is well known that the derivative of a function f in the Euclidean space at x_0 is a measure of the stability of $f(x)$ at the neighbourhood of the point x_0. For the linear complexity function $L(x)$ in the Hamming space $(GF(q)^n, d_H)$, a similar measure of the stability of $L(x)$ is

$$k(u,x) = \max_{y\in O(x,u)} |L(x) - L(y)|$$

where $k(u,x)$ resembles the derivative of a function in the Euclidean space. From the cryptographic viewpoints what we are concerned with is how much the linear complexity of a sequence decreases after changing a number of components in the sequence, but not how much it increases. Thus the indexes $WC_u(x)$, $SC_u(x)$ and $|L(x)-SC_u(x)|$ are measures of the linear-complexity stability of the sequence x.

We now list some basic properties of weight complexity for finite sequences. Let x, y be two binary sequences, then

    a) $WC_0(x)=L(x)$

    b) $WC_{n-W_H(x)}(x)\leqslant L(x)+1$

    c) $L(x)-1\leqslant WC_n(x)\leqslant L(x)+1$

    d) $WC_u(x\oplus y)\leqslant WC_u(x)+L(y)$

Proof: a) is obviously true, and b) and c) both are results of Lemma 3. 5.

    d) Noticing that $L(x+y)\leqslant L(x)+L(y)$, we get

$$WC_u(x \oplus y) = \min_{W_H(z)=u} L(x \oplus y \oplus z)$$

$$\leqslant \min_{W_H(z)=u} (L(x \oplus z) + L(y))$$

$$= WC_u(x) + L(y)$$

Let s^∞ be a sequence of period p. It is well known that $L(s^\infty)=L$ (s^{2p}), where s^N denotes the finite sequence $s_0 s_1 \cdots s_{N-1}$. To measure the stability of linear complexity of periodic sequences, we define the weight complexity or sphere surface complexity and sphere complexity respectively as in the following definition.

Definition 4. 10 Let s^∞ be a sequence of period p over $GF(q)$. Define the weight complexity and sphere complexity respectively as

$$WC_u(s^\infty) = \min_{\substack{W_H(t^p)=u \\ Per(t^\infty)=p}} L(s^\infty + t^\infty)$$

and

$$SC_u(s^\infty) = \min_{\substack{0<W_H(t^p)\leqslant u \\ Per(t^\infty)=p}} L(s^\infty + t^\infty)$$

where $per(t^\infty)$ denotes the period of t^∞.

R. A. Rueppel investigated the dynamic behaviour of the linear complexity of binary sequences. i. e. , the growth process of $L(s^n)$ with increasing n, and concluded that a good random sequence generator should have linear complexity close to the period length, and also a linear complexity profile which follows closely, but "irregularly", the $\frac{n}{2}$-line (where n denotes the number of sequence digits)[Ruep 86]. We are now not clear about the relationship between the weight complexity and sphere complexity and the linear complexity profile. But it seems that the binary sequences which have good linear complexity stability lie in the class of sequences with linear complexity profile which follows closely the $\frac{n}{2}$-line. The stability of linear complexity of periodic sequences will be investigated in detail in the later chapters.

Note: We were told by the reviewer of the monograph that Professor Harriet Fell successfully used techniques of ergodic theory to handle stability. It is a pity that we have not got in touch with her at the time of writing, although we would like to have introduced her work in the monograph.

4. 4 On the Security of Several Kinds of Key Stream Generators

In this section, we would like to investigate the security of several kinds of binary key stream generators. All the discussions are in $GF(2)$. We first give an upper bound for the sphere complexity of the nonlinear state filtered ML-sequences.

Theorem 4. 11   Assume that the feedback polynomial of the driving LFSR of the stream cipher depicted in Fig. 3. 1 is primitive and of degree n. Let s^∞ be the key stream and

$$\{ |S_{(f)}(w)| : \quad w \in GF(2)^n \} = \{V_i/2^n, \quad V_{i+1} \neq V_i\}$$

then

$$SC_{2^{n-1}-V_i/2}(s^\infty) \leqslant n + 1$$

Proof: Let t^∞ be the output sequence of the sequence generator in Fig. 4. 3.

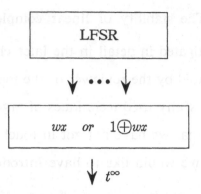

Fig. 4. 3   The sequence generator used to approximate the original one

Where the LFSR in Fig. 4. 3 is the same as the original one. Since the feedback polynomial of the driving LFSR is primitive and of degree n, the key stream has $2^n - 1$ as one of its periods (it may not be the minimal period), and the state vector takes each vector of $GF(2)^n - \{0\}$ with equal likelihood. On the other hand, the number of agreements between $f(x)$ and wx or $1 \oplus wx$ is, as x runs over $GF(2)^n$,

$$\left[\frac{1}{2} + \frac{1}{2}S_{(f)}(w)\right]2^n = 2^{n-1} + \frac{V_i}{2}.$$

Therefore the number of disagreements between $f(x)$ and wx or $1 \oplus wx$ is, as x runs over $GF(2)^n - \{0\}$,

$$2^{n-1} - \frac{V_i}{2} \text{ or } 2^{n-1} - \frac{V_i}{2} - 1.$$

Thus, the number of disagreements between $s^{2^n - 1}$ and $t^{2^n - 1}$ is less than or equal to $2^{n-1} - \frac{V_i}{2}$, and

$$SC_{2^{n-1} - V_i/2}(s^\infty) \leqslant L(t^\infty)$$

By Lemmas 3. 4 and 3. 5 we obtain $L(t^\infty) \leqslant n + 1$. Hence we have

$$SC_{2^{n-1} - V_i/2}(s^\infty) \leqslant n + 1$$

Theorem 4. 12   Assume the feedback polynomials of the n driving LFSRs of the stream cipher depicted in Fig. 3. 2 are primitive and of degree $L_1, L_2, \cdots, L_n$ respectively, and

$$max_{w \in GF(2)^*} |S_{(f)}(w)| = a$$

setting

$$B = \{w : |S_{(f)}(w)| = a\}$$

$$L = min_{w \in B}(\sum_{i=1}^{n} w_i L_i) + 1$$

then the key stream of the stream cipher can be approximated by a sequence with linear complexity less than or equal to L, and the probability of agreements is approximately $\frac{1}{2} + \frac{1}{2}a$.

Proof: Let $v \in B$ such that

$$L = \sum_{i=1}^{n} v_i L_i + 1$$

By Lemmas 3. 4 and 3. 5 we know that the linear complexity of the output sequence of the running key generator is less than or equal to L if we replace the combining function f with vx or $1 \oplus vx$. If $x = (x_1, x_2, \cdots, x_n)$ takes each vector of $GF(2)^n$ with equal likelihood, the probability of agreement between $f(x)$ and vx, or $1 \oplus vx$ is $\frac{1}{2} + \frac{1}{2}S_{(f)}(v)$. If x_i takes the output digits of the LFSR$_i$, noticing the feedback polynomial of each LFSR is primitive, we have

$$P(x_i = 1) = \frac{1}{2} + 1/(2^{L_i} - 1)$$

and we can regard $P(x_i = 1) = \frac{1}{2}$ because $2^{L_i} - 1$ is practically very large. On the other hand, the input variables $x_1, \cdots, x_n$ are statistically independent, and $L_1 + \cdots + L_n$ is much larger than n in practice. Thus the probability of agreement between the key stream and the output sequence of the sequence generator in Fig. 4. 4 is approximately $\frac{1}{2} + \frac{1}{2}a$

It follows from Lemmas 3. 4 and 3. 5 that $L(t^\infty) \leqslant L$. This proves Theorem 4. 12.   #

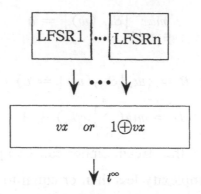

Fig 4. 4 The sequence generator used to approximate the original one

In what follows in this section, we shall analyze the security of a kind of finite-state-machine stream cipher with memory. Siegenthaler, Xiao and Massey investigated the correlation immunity of nonlinear combiners, but always under the assumption that the combiner is memoryless. Rueppel [Ruep 86] showed that for a finite-state-machine combiner it was possible to obtain maximum correlation immunity together with maximum nonlinear order. Thus the undesired trade-off could be broken through the use of memory. He also concluded that one bit of memory surfaces to realize a binary FSM-combining with maximum correlation-immune order $N-1$ and with maximum nonlinear order N (see Fig. 4. 5). For this case the FSM equations may be written as

$$z_j = \sum_{i=1}^{N} x_{ij} + s_j \qquad \qquad in\ GF(2)$$

$$s_{j+1} = f_s(x_{1j}, x_{2j}, \cdots, x_{Nj}, s_j) \qquad in\ GF(2)$$

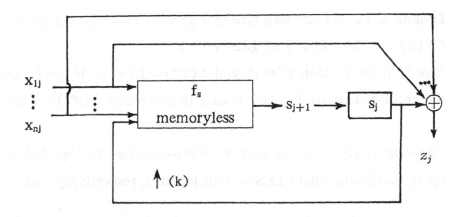

Fig. 4.5 1-bit memory FSM—combiner of maximum correlation-immunity $N-1$ allowing maximum nonlinear order in f.

Before investigating the linear-complexity stability of the above key stream generator, we need the help of the following two lemmas:

Lemma 4.13 Assume x^∞ is a binary sequence with linear complexity L. Let

$$y_i = \sum_{j=0}^{i} x_j \qquad in\ GF(2)$$

Then $\qquad L(y^\infty) \leqslant L + 1$

Proof: Suppose $(f(x), L)$ is the shortest LFSR that generates the sequence x^∞, and

$$f(x) = 1 \oplus c_1 x \oplus \cdots \oplus c_L x^L$$

By definition we have

$$x_{n+1} = c_1 x_n \oplus c_2 x_{n-1} \oplus \cdots \oplus c_L x_{n-L+1} - 0, \qquad n \geqslant L \mid 1.$$

Noticing that $\qquad x_{n+1} = y_{n+1} \oplus y_n$, we obtain

$$y_{n+1} = (1 \oplus c_1) y_n \oplus \cdots \oplus (c_{L-1} \oplus c_L) y_{i-L+1} \oplus c_L y_{i-L}, \qquad i \geqslant L$$

Hence $\qquad L(y^\infty) \leqslant L + 1$.

Lemma 4. 14   Let x^∞ be a binary sequence, and $z_{j+1} = x_j \oplus cz_j$, $c \in GF(2)$. Then $L(z^\infty) \leqslant L(x^\infty) + 1$.

Proof: If $c = 0$, then $z^\infty = x^\infty$ and $L(z^\infty) = L(x^\infty)$. If $c = 1$, similar to the proof of Lemma 4. 13, it is easy to prove that $L(z^\infty) \leqslant L(x^\infty) \oplus 1$.

Theorem 4. 15   Assume that the FSM-combiner in Fig. 4. 5 is driven by N maximum length LFSRs with length n_i respectively, let

$$a = \max_w |S_{(f)}(w)|$$

and

$$B = \{w: |S_{(f)}(w)| = a\}$$

Then there exists a sequence with linear complexity L such that

$$L \leqslant \sum_{i=1}^{N} n_i + \min_{w \in B}(\sum_{i=1}^{N} w_i n_i) + 2$$

and with the probability of agreement with the original one approximately $\dfrac{1}{2} + \dfrac{1}{2}a$.

Proof: Let $v \in B$ satisfy

$$\sum_{i=1}^{N} v_i n_i = \min_{w \in B}(\sum_{i=1}^{N} w_i n_i)$$

We now construct a new FSM-combiner which is driven by the same N maximum length LFSRs. The next state function and output function are respectively

$$y_j = \sum_{i=1}^{N} x_{ij} + t_j \qquad in \ GF(2)$$

$$t_{j+1} = \sum_{i=1}^{N} v_i x_{ij} + v_{N+1} t_j + e \qquad in \ GF(2)$$

where $e = 1$ if $S_{(f)}(v) < 0$, $e = 0$ if $S_{(f)}(v) \geqslant 0$.

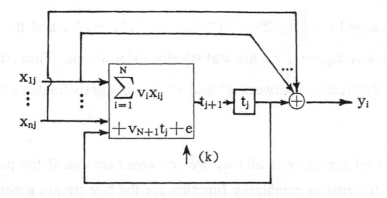

Fig. 4. 6 The constructed FSM-combiner to approximate the original one

Let

$$c_i = \sum_{i=1}^{N} v_i x_{ij} \qquad in\ GF(2)$$

$$d_i = \sum_{i=1}^{N} x_{ij} \qquad in\ GF(2)$$

It follows from Lemma 3. 4 and 3. 5 that

$$L(c^\infty) \leqslant \sum_{i=1}^{N} v_i n_i$$

$$L(d^\infty) \leqslant \sum_{i=1}^{N} n_i$$

It followes from Lemma 4. 14 that $L(t^\infty) \leqslant L(d^\infty) + 2$.

Therefore we have

$$L(y^\infty) \leqslant L(d^\infty) + L(t^\infty)$$

$$\leqslant \sum_{i=1}^{N} n_i + \sum_{i=1}^{N} v_i n_i + 2$$

$$= \sum_{i=1}^{N} n_i + \min_{w \in B} (\sum_{i=1}^{N} w_i n_i) + 2$$

Since the feedback polynomials of the N maximum length LFSRs are primitive, $P(x_{ij}=1)=\dfrac{1}{2}+(2^{L_i}-1)^{-1}$ for $i=1,2,\cdots,N$, and the N input varibles $x_{1j}, x_{2j},\cdots, x_{Nj}$ are statistically independent. Thus, the probabilility of agreement between z^{∞} and y^{∞} is well approximated by $\dfrac{1}{2}$ $+\dfrac{1}{2}a$.

Based on all the analyses above, we can conclude that if the peak values of the filtering or combining function for the key stream generators in Figs. 3. 1 and 3. 2 and of the next state function f_s for the FSM-combiner in Fig. 4. 6 are large enough, the key stream generators are not secure under the known plaintext-ciphertext pair attack. This is due to the fact that the stability of linear complexity of the key streams is not good in these key stream generators. In what follows in this section, we shall give a lower bound for the weight complexities of binary de Bruijn sequences.

Theorem 4. 16 Let s^{∞} be a binary de Bruijin sequence of period 2^n, which is generated by a nonlinear feedback shift register with feedback function $f(x_1,x_2,\cdots, x_n)$. Then for each $w\in GF(2)^n$,

$$WC_{k(w)}(s^{\infty}) \leqslant n+1$$

where $\quad k(w) = 2^{n-1}(1 - |S_{(f)}(w)|)$

Proof: It is well known that the state vector of a nonlinear feedback shift register (NFSR) takes each vector of $GF(2)^n$ once and only once while the NFSR produces one period of a de Bruijn sequence of period 2^n. Thus, the number of agreements between the first period of the output sequence of the NFSR and that of the linear feedback shift register with the same initial state vector and the feedback function wx or $1\oplus wx$, is $2^n(\dfrac{1}{2}+\dfrac{1}{2}|S_{(f)}(w)|)$. Therefore, the number of disagreements is $2^n - 2^{n-1} - 2^{n-1}|S_{(f)}(w)| = k(w)$. Hence, Theorem 4. 16 holds.

It is known that if s^∞ is a de Bruijn sequence of period $2^n (n \geqslant 3)$, then $2^{n-1} + n \leqslant L(s^\infty) \leqslant 2^n - 1$ [Chan 82]. It follows from Theorem 4. 16 and the above result that a de Bruijn sequence cannot be used as a running key sequence for stream ciphers if the peak value of the spectra of the nonlinear feedback function of the NFSR that generates it is large enough.

4. 5 On the Stability of Elementary Symmetric Boolean Functions

Integer addition has been employed in the knapsack public key cryptosystems and the knapsack stream ciphers [Diff 76] [Ruep 86]. It has been shown that integer addition is a cryptographically useful function, since its $GF(2)$-interpretation has high nonlinear order. Let y be the integer sum of N binary variables $x_1, \cdots, x_N$, i. e. ,

$$y = \sum_{i=1}^{N} x_i$$

and let

$$y = y_0 + y_1 2 + y_2 2^2 + \cdots + y_r 2^r$$

be the binary representation of y. Then

$$y_i = f_{2^i}(x_1, x_2, \cdots, x_N)$$

where

$$f_k(x_1, x_2, \cdots, x_N) = \sum_{1 \leqslant i_1 < \cdots < i_k \leqslant N} x_{i_1} x_{i_2} \cdots x_{i_k}$$

denotes the $GF(2)$-sum of all distinct kth-order products of N binary variables and $r=[log_2 N]$. Although $f_{2^i}(x_1, x_2, \cdots, x_N)$ may have large nonlinear order, its stability may be bad. since $PV(f_1)=max|S_{(f_1)}(w)|=1$, the stability of $f_1(x)$ is the worst. This deserves some consideration. For example, the following stream cipher is based on integer addition. Let us see how to break the stream cipher by a known plaintext-ciphertext-pair attack.

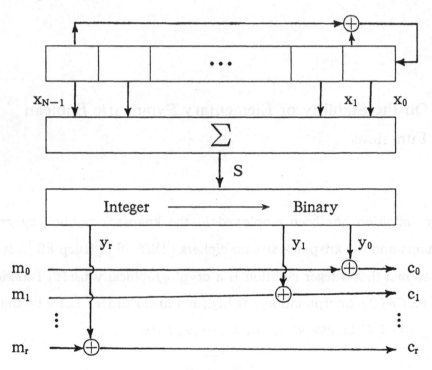

Fig. 4. 7 A stream cipher based on integer additon

Suppose a cryptanalyst only know the structure of the stream cipher, i. e. , that it consists of two parts, the driving LFSR and the module operation. This means the keys of the stream cipher comprise the initial state and the feedback polynomial of the driving LFSR. Then the stream cipher can be broken provided that $2([log_2 N]+1)N$ consecutive bits of plaintext-ciphertext pairs are known to the cryptanalyst. Let $s_0 s_1 \cdots s_{k-1}$, $k = 2([log_2 N] + 1)N$, be the k consecutive bits of the key stream, and

$$t_i^{2N} = s_i s_{M-1+i} s_{2(M-1)+i} \cdots s_{2(N-1)(M-1)+i} \qquad \text{for } 0 \leqslant i \leqslant M - 1.$$

Then use the Berlekamp-Massey algorithm to find the shortest LFSR that generates t_i^{2N} for $i = 0, 1, \cdots, M-1$, say, $(g_0(x), L_0), \cdots, (g_{M-1}(x), L_{M-1})$. Let $L_p = min(l_0, \cdots, L_{M-1})$, we can expect t_p^{2N} to be the $2N$ consecutive bits generated by the 0-th output stage. Whence $g_p(x)$ is the feedback polynomial of the driving LFSR. By using the method in Section 3. 3, we can find the corresponding state vector of the driving LF-SR by which $s_0 s_1 \cdots s_{M-1}$ is produced. After this process, use $t_i^{2N} (i \neq p)$ to check whether the calculated feedback polynomial and the corresponding state vector is the original. If not, let

$$L_q = min(L_0, \cdots, L_{p-1}, L_{p+1}, \cdots, L_{M-1})$$

and repeat the above process untill we find the true one.

That the above attack is successful is due to the fact that $f_1(x_1, x_2, \cdots, x_N)$ is not stable. To employ integer addition in stream ciphers, it is necessary for us to make clear the stability of $f_{2^i}(x_1, x_2, \cdots, x_N)$, which belongs to the class of elementary symmetric Boolean functions. Functions such as those with the form of $f_k(x_1, x_2, \cdots, x_N)$ defined at the beginning of this section are called elementary symmetric Boolean functions. In Section 4. 2 we have already proved that $f_2(x_1, x_2, x_N)$ is bent if N is even. Thus $PV(f_2) = 2^{-\frac{N}{2}}$. For simplicity of expression, we stipulate that $f_K^N(x)$ denotes $f_K(x_1, \cdots, x_N)$.

Theorem 4. 17   For odd N, we have

$$S_{(f_2^N)}(w) = \pm 2^{-\frac{N-3}{2}} \text{ or } 0$$

Proof : It is easy to check that

$$f_2^N(x_1, x_2, \cdots, x_N) = x_1 f_1^{N-1}(x_2, x_3 \cdots, x_N) + f_2^{N-1}(x_2, x_3, \cdots, x_N)$$

It follows that

$$S_{(f_2^N)}(w_1, \cdots, w_N) = \sum_{(x_1, \cdots, x_N) \in GF(2)^N} (-1)^{x_1 f_1^{N-1}(x_2, \cdots, x_N) + f_2^{N-1}(x_2, \cdots, x_N) + \sum_{i=1}^{N} w_i x_i}$$

$$= \sum_{(x_2, \cdots, x_N) \in GF(2)^{N-1}} (-1)^{f_2^{N-1}(x_2, x_3, \cdots, x_N) + \sum_{i=1}^{N} w_i x_i}$$

$$+ \sum_{(x_2, \cdots, x_N) \in GF(2)^{N-1}} (-1)^{f_2^{N-1}(x_2, \cdots, x_N) + \sum_{i=1}^{N} \bar{w}_i x_i + w_1}$$

$$= S_{(f_2^{N-1})}(w_2, \cdots, w_N) + (-1)^{w_1} S_{(f_2^{N-1})}(\bar{w}_2, \cdots, \bar{w}_N)$$

Noticing that f_2^{N-1} is bent and denoting $S_{(f_2^{N-1})}(w) = (-1)^{e(w)} 2^{-\frac{N}{2}}$, we get

$$S_{(f_2^N)}(w_1, w_2, \cdots, w_N) = (-1)^{e(w_2, w_3, \cdots, w_N)} + (-1)^{e(\bar{w}_1, \cdots, \bar{w}_N) + w_1}$$

Thus,

$$|S_{(f_2^N)}(w)| = 2^{-\frac{N-3}{2}} \text{ or } 0 \qquad\qquad \#$$

So far we have proved that for any elementary symmetric function $f_2^N(x_1, x_2, \cdots, x_N)$, we have

$$PV(f_2^N(x)) \leqslant 2^{-\frac{N}{2} + \frac{3}{2}}$$

Thus they are relatively stable; unfortunately, their nonlinear orders are only 2. Generally, we have the following conclusion about the spectra of elementary symmetric functions.

Theorem 4. 18   For any $w \in GF(2)^N$, it holds that

$$S_{(f_x^N)}(w) = \frac{1}{2^N} \sum_{j=1}^{N} (-1)^{\binom{j}{i}} \sum_{h \leqslant t \leqslant m} (-1)^i \begin{bmatrix} l \\ t \end{bmatrix} \begin{bmatrix} N - l \\ j - t \end{bmatrix}$$

where $l = W_H(w)$, $h = max\{j + l - N, 0\}$ and $m = min\{j, l\}$.

Proof: By definition, it is easy to see that $S_{(f_i)}(w) = S_{(f_i)}(v)$ provided that $W_H(w) = W_H(v)$. Therefore, let us consider $S_{(f_i^N)}(w)$ with $w = (\underbrace{1 \ 1 \ \cdots \ 1}_{l} \ 0 \ 0 \ \cdots \ 0)$ for $0 \leqslant l \leqslant N$. Noticing that

$$f_k^N(x) = \begin{cases} 0 & , & W_H(x) < k \\ \left\lfloor \dfrac{W_H(x)}{k} \right\rfloor (\bmod\ 2), & W_H(x) \geq k \end{cases}$$

and stipulating that $\begin{vmatrix} p \\ q \end{vmatrix} = 0$ for $p < q$, we have

$$S_{(f_k)}(w) = \frac{1}{2^N} \sum_x (-1)^{\binom{W_H(x)}{k} + wx}$$

$$= \frac{1}{2^N} \sum_{j=0}^{N} (-1)^{\binom{j}{k}} \sum_{W_H(x)=j} (-1)^{wx}$$

For any $x = (x_1, x_2, \cdots, x_N)$ with $W_H(x) = j$, let $t = W_H((x_1, x_2, \cdots, x_l))$, then $j - t = W_H((x_{l+1}, \cdots, x_N))$. It follows that

$$h = max\{j + l - N,\ 0\} \leqslant t \leqslant min\{j,\ l\} = m$$

Thus we get

$$S_{(f_k^N)}(w) = \frac{1}{2^N} \sum_{j=0}^{N} (-1)^{\binom{j}{k}} \sum_{h \leqslant t \leqslant m} (-1)^t \begin{vmatrix} l \\ t \end{vmatrix} \begin{vmatrix} n - l \\ j - t \end{vmatrix}$$

In order to analyze the stability of $f_k^N(x)$, one has to simplify the above formula. Unfortunately, we have not found ways to simplify the above combinatorial formula. Therefore, based on Theorem 4.18 and with the aid of computers, we have calculated the spectra of some elementary symmetric Boolean functions. We present the results for reference. Since the spectra of $f_k^N(x)$ have already been formulated for $k \leqslant 2$, we only list $f_k^N(x)$ for $k \geqslant 3$.

Table 4. 1 Spectra of $f_k^6(x)$ for $k \geqslant 3$

k	$W_H(w)$	$S_{(f_k^6)}(w)$	k	$W_H(w)$	$S_{(f_k^6)}(w)$
3	0	.3750	5	0	.8125
3	1	.0000	5	1	.1250

k	$W_H(w)$	$S_{(f_k^6)}(w)$	k	$W_H(w)$	$S_{(f_i^6)}(w)$
3	2	.1250	5	2	$-.0625$
3	3	.0000	5	3	.0000
3	4	$-.1250$	5	4	.0625
3	5	.0000	5	5	$-.1250$
3	6	.6250	5	6	.1875
4	0	.3125	6	0	.9688
4	1	.3125	6	1	.0313
4	2	.0625	6	2	$-.0313$
4	3	$-.0625$	6	3	.0313
4	4	.0625	6	4	$-.0313$
4	5	.0625	6	5	.0313
4	6	$-.3125$	6	6	$-.0313$

Table 4. 2   Spectra of $f_k^7(x)$ for $k \geqslant 3$

k	$w_H(w)$	$S_{(f_k^7)}(w)$	k	$w_H(w)$	$S_{(f_k^7)}(w)$
3	0	.4375	5	4	.0313
3	1	−.0625	5	5	.0313
3	2	.0625	5	6	−.1563
3	3	.0625	5	7	.3438
3	4	−.0625	6	0	.8750
3	5	−.0625	6	1	.0938
3	6	.0625	6	2	−.0625
3	7	.5625	6	3	.0313
4	0	.0000	6	4	.0000
4	1	.3125	6	5	−.0313
4	2	.0000	6	6	.0625
4	3	−.0625	6	7	−.0938
4	4	.0000	7	0	.9844
4	5	.0625	7	1	.0156
4	6	.0000	7	2	−.0156
4	7	−.3125	7	3	.0156
5	0	.6563	7	4	−.0156
5	1	.1563	7	5	.0156

k	$w_H(w)$	$S_{(f_k^7)}(w)$	k	$w_H(w)$	$S_{(f_k^7)}(w)$
5	2	$-.0313$	7	6	$-.0156$
5	3	$-.0313$	7	7	$.0156$

Table 4. 3 Spectra of f_k^8 for $k \geqslant 3$

k	$W_H(w)$	$S_{(f_k^8)}(w)$	k	$W_H(w)$	$S_{(f_k^8)}(w)$
3	0	$.5000$	6	0	$.7188$
3	1	$-.0625$	6	1	$.1563$
3	2	$.0000$	6	2	$-.0625$
3	3	$.0625$	6	3	$.0000$
3	4	$.0000$	6	4	$.0313$
3	5	$-.0625$	6	5	$-.0313$
3	6	$.0000$	6	6	$.0000$
3	7	$.0625$	6	7	$.0625$
3	8	$.0500$	6	8	$-.1563$
4	0	$-.2656$	7	0	$.9375$
4	1	$.2656$	7	1	$.0469$
4	2	$.0469$	7	2	$-.0313$
4	3	$-.0469$	7	3	$.0156$
4	4	$-.0156$	7	4	$.0000$

k	$W_H(w)$	$S_{(f_4^8)}(w)$	k	$W_H(w)$	$S_{(f_4^8)}(w)$
4	5	.0156	7	5	$-.0156$
4	6	.0469	7	6	.0313
4	7	.0469	7	7	$-.0469$
4	8	$-.2656$	7	8	.0625
5	0	.5000	8	0	.9922
5	1	.1563	8	1	.0078
5	2	.0000	8	2	$-.0078$
5	3	$-.0313$	8	3	.0078
5	4	.0000	8	4	$-.0078$
5	5	.0313	8	5	.0078
5	6	.0000	8	6	$-.0078$
5	7	$-.1563$	8	7	.0078
5	8	.5000	8	8	$-.0078$

From our calculations of the spectra of elementary symmetric functions $f_K^N(x)$ for $N=5,6,\cdots,15$, we find that a lot of them have good stability. But there do exist many among them which are not stable. We would like to mention here that although $PV(f)$ and $VS(f)$, defined in Section 4.1, are measure indexes for the stability of Boolean functions. $VS(f)$ is relatively more suitable. Obviously, $VS(f)=0$ iff $f(x)$ is a bent function. Due to the fact that there are many elementary symmetric Boolean functions which are not stable, although integer addition may, as a whole, has good properties and behaves well, one has to be careful while employing it in stream ciphers, especially employing its GF(2)-interpretation functions separately.

Although the stability of some elementary symmetric functions is bad, there do exist some symmetric Boolean functions with ideal stability. A Boolean function $f(x_1, x_2, \cdots, x_n)$ is said to be symmetric if $f(x_1, x_2, \cdots, x_n)$ is fixed under the permutation group of n symbols, i. e. , for every permutation $P: i \to i'$ of $\{1, 2, \cdots, n\}$, $f(x_1, x_2, \cdots, x_n) = f(x_{P(1)}, x_{P(2)}, \cdots, x_{P(n)})$. It is well known from elementary algebra that every symmetric function is expressible as a function in the elementary symmetric functions. Strict majority logic functions are symmetric ones with ideal stability, which are used for the fast synchronization of sequences [Tits 64]. Based on SML functions, Bruer proposed the following SML generator [Brue 84][Sieg 86], and suggested employing it as a running-key generator (Fig. 4. 8).

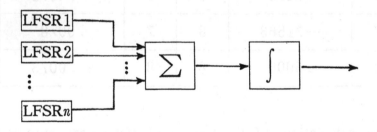

Fig. 4. 8 The SML generator

Siegenthaler estimated the number of ciphertext digits needed to attack the generator by the correlation approach [Sieg 86]. In order to analyze the suitablity of SML functions as combining ones, we need to know their stability. Let n be an odd positive number; the function of n

arguments is defined as

$$g(x) = \begin{cases} 1, & if \ W_H(x) > \dfrac{n}{2} \\ 0, & if \ W_H(x) < \dfrac{n}{2} \end{cases}$$

Let us now calculate the spectra of $g(x)$. Noticing that n is odd, we get

$$\sum_{W_H(x) > \frac{n}{2}} (-1)^{wx} = \sum_{W_H(y) < \frac{n}{2}} (-1)^{w(y+1)}$$

$$= (-1)^{W_H(w)} \sum_{W_H(y) < \frac{n}{2}} (-1)^{wy}$$

where $1 = (1, 1, \cdots, 1)$. Whence we get

$$S_{(g)}(w) = 2^{-n} \sum_{x \in GF(2)^n} (-1)^{g(x) + wx}$$

$$= \begin{cases} 0, & if \ W_H(w) \ is \ even \\ 2^{-n+1} \displaystyle\sum_{W_H(x) < \frac{n}{2}} (-1)^{wx}, & if \ W_H(w) \ is \ odd. \end{cases}$$

Thus we have proved the following result:

Theorem 4. 19 Let n be odd, and $g(x_1, x_2, \cdots, x_n)$ be the SML function, then $S_{(g)}(w) = 0$ for each w with $W_H(w)$ being even.

Similar to the proof of Theorem 4. 18, we can prove that

$$S_{(g)}(w) = 2^{-n+1} \sum_{j=0}^{(n-1)/2} \sum_{h \leqslant l \leqslant m} (-1)^t \binom{l}{t} \binom{n-l}{j-t}$$

where $h = max\{j+l-n, 0\}$, and $m = min\{j, l\}$. Especially, for w with $W_H(w) = 1$, we have

$$S_{(g)}(w) = 2^{-n+1} \left[\sum_{i=1}^{(n-1)/2} \left[\binom{n}{i} - \binom{n-1}{i-1} \right] - \sum_{i=0}^{(n-3)/2} \binom{n-1}{i} + \binom{n}{0} \right]$$

$$= \binom{n-1}{(n-1)/2} / 2^{n-1}$$

For $w = (1, 1, \cdots, 1)$, we have

$$S_{(g)}(w) = 2^{-n+1} \sum_{W_H(x) < \frac{n}{2}} (-1)^{wx}$$

$$= 2^{-n+1} \sum_{i=0}^{\frac{n-1}{2}} (-1)^i \binom{n}{i}$$

$$= (-1)^{\frac{n-1}{2}} \binom{n-1}{\frac{n-1}{2}} 2^{-n+1}$$

Generally, for w with $W_H(w) = l \leqslant \dfrac{n-1}{2}$ and l odd, we get

$$S_{(g)}(w) = \sum_{i=0}^{l} (-1)^i \binom{l}{i} \sum_{j=0}^{\frac{n-1}{2}-i} \binom{n-l}{j}$$

$$= \sum_{j=0}^{\frac{n-1}{2}-l} \binom{n-l}{j} \sum_{i=0}^{l} (-1)^i \binom{l}{i}$$

$$+ \sum_{i=0}^{l-1} \binom{n-l}{\frac{n-1}{2}-i} \sum_{j=0}^{i} (-1)^j \binom{l}{j}$$

Noticing that

$$\sum_{i=0}^{k} \binom{m}{i} (-1)^i = (-1)^k \binom{m-1}{k}$$

and

$$\sum_{i=0}^{l} (-1)^i \binom{l}{i} = 0$$

we obtain

$$S_{(g)}(w) = \sum_{i=0}^{l-1} \binom{n-l}{\frac{n-1}{2}-i} \binom{l-1}{i} (-1)^i$$

On the other hand, we have

$$\begin{bmatrix} n-l \\ \dfrac{n-1}{2}-i \end{bmatrix}\begin{bmatrix} l-1 \\ i \end{bmatrix} = \frac{(n-l)!\,(l-1)!}{\left[\left|\dfrac{n-1}{2}\right|!\right]^2}\begin{bmatrix} \dfrac{n-1}{2} \\ l-1-i \end{bmatrix}\begin{bmatrix} \dfrac{n-1}{2} \\ i \end{bmatrix}$$

Whence

$$S_{(g)}(w) = \frac{(n-l)!\,(l-1)!}{\left[\left|\dfrac{n-1}{2}\right|!\right]^2}\left[\sum_{i=0}^{l-1}\begin{bmatrix} \dfrac{n-1}{2} \\ l-1-i \end{bmatrix}\begin{bmatrix} \dfrac{n-1}{2} \\ i \end{bmatrix}(-1)^i\right].$$

Noticing that

$$\left[\sum_{j=0}^{\frac{n-1}{2}}\begin{bmatrix} n \\ j \end{bmatrix}x^j\right]\left[\sum_{j=0}^{\frac{n-1}{2}}(-1)^j\begin{bmatrix} n \\ j \end{bmatrix}x^j\right]$$

$$= (1+x)^{\frac{n-1}{2}}(1-x)^{\frac{n-1}{2}}$$

$$= (1-x^2)^{\frac{n-1}{2}}$$

$$= \sum_{i=0}^{\frac{n-1}{2}}(-1)^i\begin{bmatrix} \dfrac{n-1}{2} \\ i \end{bmatrix}x^{2i}$$

we get

$$(-1)^{\frac{l-1}{2}}\begin{bmatrix} \dfrac{n-1}{2} \\ \dfrac{l-1}{2} \end{bmatrix} = \sum_{i=0}^{l-1}\begin{bmatrix} \dfrac{n-1}{2} \\ l-1-i \end{bmatrix}\begin{bmatrix} \dfrac{n-1}{2} \\ i \end{bmatrix}.$$

Therefore,

$$S_{(g)}(w) = (-1)^{\frac{l-1}{2}}\frac{(n-l)!\,(l-1)!}{(\dfrac{n-1}{2})!\,(\dfrac{n-l}{2})!\,(\dfrac{l-1}{2})!}$$

Similarly, for w with $W_H(w) = l \geqslant \dfrac{n+1}{2}$, we have

$$S_{(g)}(w) = \sum_{i=0}^{n-l}\begin{bmatrix} n-l \\ i \end{bmatrix}\begin{bmatrix} l-1 \\ \dfrac{n-1}{2}-i \end{bmatrix}(-1)^{\frac{n-1}{2}-i}$$

setting $n - l = L - 1$, we obtain

$$S_{(g)}(w) = (-1)^{\frac{n-1}{2}} \sum_{i=0}^{L-1} \begin{bmatrix} L-1 \\ i \end{bmatrix} \begin{bmatrix} n-L \\ \dfrac{n-1}{2} - i \end{bmatrix} (-1)^i$$

By the result proved above , we have

$$S_{(g)}(w) = (-1)^{\frac{n-1}{2}+\frac{L-1}{2}} \frac{(n-L)!(L-1)!}{(\dfrac{n-1}{2})!(\dfrac{n-L}{2})!(\dfrac{L-1}{2})!}$$

$$= (-1)^{n-\frac{l+1}{2}} \frac{(l-1)!(n-l)!}{(\dfrac{n-1}{2})!(\dfrac{l-1}{2})!(\dfrac{n-l}{2})!}$$

Summarizing the foregoing results, for each w with $W_H(w)$ odd we have

$$|S_{(g)}(w)| = \frac{(l-1)!(n-l)!}{(\dfrac{n-1}{2})!(\dfrac{l-1}{2})!(\dfrac{n-l}{2})!}$$

Note: The above result for $S_{(g)}(w)$ was obtained by Titsworth [Tits 64] but we have derived it here with a simple method.

Let a_l denote the right side of the above formula, then we get

$$\frac{a_l + 2}{a_l} = \frac{l}{n - l - 1}$$

Thus we obtain $a_{l+2} \leqslant a_l$ for $l \leqslant \dfrac{n-1}{2}$, and $a_{l+2} \geqslant a_l$ for $l \geqslant \dfrac{n+1}{2}$.

So far we have proved the following conclusion.

Theorem 4. 20 Let n be odd, and $g(x_1, x_2, \cdots, x_n)$ the SML function of n arguments, then

$$\max_{w \in GF(2)^*} |S_{(g)}(w)| = \begin{bmatrix} n-1 \\ \dfrac{n-1}{2} \end{bmatrix} / 2^{n-1}$$

To analyze the cryptographic properties of SML functions, we need to know the dynamic behavior of $\begin{bmatrix} n-1 \\ \dfrac{n-1}{2} \end{bmatrix} / 2^{n-1}$, denoted here as

b_{n-1}. Noticing

$$\frac{b_{n-1}}{b_{n-3}} = \frac{n-3}{n-2}$$

we get

$$b_{n-1} = \left[\frac{b_5}{b_3} \cdot \frac{b_7}{b_5} \cdots \frac{b_{n-1}}{b_{n-3}}\right] b_3$$

$$= \frac{3 \cdot 5 \cdot 7 \cdots (n-3)}{4 \cdot 6 \cdot 8 \ldots (n-2)} b_3$$

setting

$$\alpha_{2k} = \frac{2 \cdot 4 \cdots (2k)}{3 \cdot 5 \cdots (2k+1)}$$

$$\beta_{2k} = \frac{3 \cdot 5 \cdots (2k+1)}{4 \cdot 6 \cdots (2k+2)}$$

$$\gamma_{2k} = \frac{4 \cdot 6 \cdots (2k+2)}{5 \cdot 7 \cdots (2k+3)}$$

and noticing that $\alpha_k < \beta_k < \gamma_k$, we have

$$\beta_{2k}^2 > \alpha_{2k}\beta_{2k} = \frac{1}{k+1}$$

$$\beta_{2k}^2 < \beta_{2k}\gamma_{2k} = \frac{3}{2k+3}$$

whence,

$$\sqrt{\frac{1}{k+1}} < \beta_{2k} < \sqrt{\frac{3}{2k+3}}$$

It follows that $\lim\limits_{n \to \infty} b_{n-1} = 0$. By summarizing the above results, we see that the stability of SML functions is relatively good for large n.

In order to analyze the cryptographic suitabilty of SML functions further, we have to estimate the nonlinear orders of them, since the linear complexity of the key stream is strongly connected with it. Let $g(x) = g(x_1, x_2, \cdots, x_n)$ be a SML function, then we have

$$g(x) = \sum_{W_H(c) \geq \frac{n+1}{2}} x^c$$

$$= \sum_{W_H(c) \leq \frac{n-1}{2}} \prod_{i=1}^{n} (x_i + c_i)$$

$$= \sum_{W_H(c) \leq \frac{n-1}{2}} x_1 x_2 \cdots x_n$$

$$+ (\sum_{k=1}^{n-1} \sum_{1 \leq i_1 < \cdots < i_k \leq n} \sum_{W_H(c) \leq \frac{n-1}{2}} c_{i_1} c_{i_2} \cdots c_{i_k}) \prod_{i \neq i_1, i_2, \cdots, i_k} x_i$$

$$+ \sum_{W_H(c) \leq \frac{n-1}{2}} c_1 c_2 \cdots c_n$$

For each $k \geq \dfrac{n+1}{2}$, we have

$$\sum_{W_H(c) \leq \frac{n-1}{2}} c_{i_1} c_{i_2} \cdots c_{i_k} = 0$$

Whence, we have proved the following conclusion.

Theorem 4. 21   Let n be odd, $g(x) = g(x_1, x_2, \cdots, x_n)$ be the SML function of n arguments, then there is no term of nonlinear order less than or equal to $\dfrac{n-1}{2}$ in its algebraic normal form.

Theorem 4. 22   Let n be odd, $g(x_1, x_2, \cdots, x_n)$ be the SML function of arguments, then $g(x_1, x_2, \cdots, x_n)$ has all terms of nonlinear order $\dfrac{n+1}{2}$.

Proof: For $k = \dfrac{n-1}{2}$, and $1 \leq i_1 < \cdots < i_k \leq n$, Since $n-k = \dfrac{n+1}{2}$, we have

$$\sum_{W_H(c) \leq \frac{n-1}{2}} c_{i_1} c_{i_2} \cdots c_{i_k} = 1$$

This proves the theorem.　　　#

Example 1　　$g(x_1, x_2, \cdots, x_7) = f_4^7(x_1, x_2, \cdots, x_7)$

$$= \sum_{1 \leqslant i < j < k < l \leqslant 7} x_i x_j x_k x_l$$

Generally, we have the following result:

Theorem 4. 23　Let $g(x_1, x_2, \cdots, x_n)$ be the SML function, and n an odd number, then $g(x)$ has terms with nonlinear order k iff

$$\varepsilon_k = \sum_{i=0}^{k - \frac{n+1}{2}} \binom{k}{i}$$

is odd, where $\dfrac{n+1}{2} \leqslant k \leqslant n-1$.

Proof: It follows from the following formula:

$$\sum_{W_H(c) \leqslant \frac{n-1}{2}} c_{i_1} c_{i_2} \cdots c_{i_{n-k}} = \sum_{i=0}^{k - \frac{n+1}{2}} \binom{k}{i} \qquad (mod \quad 2) \qquad \#$$

Theorem 4. 24　Let $g(x_1, x_2, \cdots, x_n)$ be the SML function, then $deg(g) \leqslant n-1$.

Proof: It follows from

$$\sum_{i=0}^{\frac{n-1}{2}} \binom{n}{i} = 2^{n-1} \qquad\qquad \#$$

Theorem 4. 25　Let $g(x_1, x_2, \cdots, x_n)$ be the SML function with $n \geqslant 3$, then there is a term of oder $n-1$ iff $\left\lceil \dfrac{n-1}{2} \right\rceil \Big/ 2$ is odd.

Proof: Noticing that

$$\varepsilon_{n-1} = \sum_{i=0}^{\frac{n-3}{2}} \binom{n-1}{i} = 2^{n-2} - \left\lceil \dfrac{n-1}{2} \right\rceil \Big/ 2$$

it follows from Theorem 4. 23 that the conclusion of Theorem 4. 25 holds.　　　#

Example 2　　$g(x_1, x_2, x_3) = x_1 x_2 + x_1 x_3 + x_2 x_3$

$$= f_2^3(x_1, x_2, x_3)$$

Lemma 4. 26　Let n be an odd number, then $\left\lfloor \dfrac{\left\lfloor \dfrac{n-1}{2} \right\rfloor}{n-1} \right\rfloor /2$ is odd iff $n = 2^m + 1$ for some positive number m.

Proof: Suppose $n = 2^m + 1$, then

$$S \overset{\triangle}{=} \left\lfloor \frac{\left\lfloor \frac{n-1}{2} \right\rfloor}{n-1} \right\rfloor /2 = \left\lfloor \frac{2^m}{2^{m-1}} \right\rfloor /2 = \prod_{i=1}^{2^{m-1}-1} \frac{2^{m-1}+i}{i}$$

Obviously, $(2^{m-1}+i)/gcd(2^{m-1}+i, i)$ is odd for each i with $1 \leqslant i \leqslant 2^{m-1}-1$, whence S is odd.

Conversely, suppose $n-1 = 2^m P$ with P odd and $P > 1$, $m \geqslant 1$, we now prove S is even. Noticing that

$$S = \left\lfloor \frac{\left\lfloor \frac{n-1}{2} \right\rfloor}{n-1} \right\rfloor /2 = \left\lfloor \frac{2^m P}{2^{m-1} P} \right\rfloor /2 = \prod_{i=1}^{2^{m-1}P-1} \frac{2^{m-1}P+i}{i}$$

let $i = 2^{j_i} P_i$ with P_i odd for $1 \leqslant i \leqslant 2^{m-1}P-1$. It is apparent that both $(2^{m-1}P+i)/2^{j_i}$ and $i/2^{j_i}$ are odd for each $j_i \neq 2^{m-1}$. Setting $i' = 2^{m-1}$ and $j_{i'} = 2^{m-1}$, then $(2^{m-1}P+i)/i = P+1$, which is even. Hence S is even. Thus P must be equal to 1. This proves the lemma.　#

It follows from Theorem 4. 25 and Lemma 4. 26 that the following conclusion holds:

Corollary 4. 27　$g(x_1, x_2, \cdots, x_n)$ has terms of order $n-1$, iff $n = 2^m + 1$ for some positive mumber m.

Theorem 4. 28　$g(x_1, x_2, \cdots, x_n)$ has all terms of order $n-2$ iff $\left\lfloor \dfrac{\left\lfloor \dfrac{n-2}{2} \right\rfloor}{n-3} \right\rfloor$

is odd, where $n \geqslant 5$.

Proof: Since $\varepsilon_{n-2} = 2^{n-3} - \begin{pmatrix} n-2 \\ n-3 \\ \hline 2 \end{pmatrix}$, it follows from Theorem 4. 23 that

Theorem 4. 28 holds. #

Corollary 4. 29 Let $n = 2^m P + 1$ with $m \geqslant 1$ and P odd.

i) If $m > 1$, then $g(x_1, x_2, \cdots, x_n)$ has all terms of order $n-2$ iff P $= 1$, i. e. , $n = 2^m + 1$;

ii) If $m = 1$, i. e. , $n = 2P + 1$ with P odd, let $P = 2^k q - 1$ with q odd. Then $g(x_1, x_2, \cdots, x_n)$ has terms of order $n-2$ iff $q = 1$.

Proof: Let

$$S = \begin{pmatrix} n - 2 \\ n - 3 \\ \hline 2 \end{pmatrix} = \prod_{i=1}^{\frac{n-3}{2}} \frac{2^{m-1}P + i - 1}{i}$$

We now consider the pairs $(2^{m-1}P + i - 1, \ i - 1)$ for $i = 3, 5, \cdots,$ $\frac{n-3}{2}$. Let $i - 1 = 2^{m_i} P_i$ with P_i odd for each i. If $m_i \neq m - 1$, then both $\frac{2^{m-1}P + i - 1}{2^{\min(m-1, m_i)}}$ and P_i are odd. On the other hand, if $P > 1$, then $i' = 2^{m-1} + 1 \leqslant \frac{n+3}{2} = 2^{m-1}P - 1$, and $(2^{m-1}P + i - 1)/2^{m-1}$ is even.

Therefore S is odd iff $P = 1$. This proves part i).

Noticing that

$$S = \prod_{i=1}^{P-1} \frac{P + i - 1}{2}$$

we can similarly prove that S is odd iff $q = 1$. #

Corollary 4. 30 $g(x_1, x_2, \cdots, x_n)$ has all terms of order $\frac{n+3}{2}$ iff $\frac{n+5}{2}$

is odd.

Proof: It follows from Theorem 4. 23 and

$$\mathcal{E}_{(n+1)/2} = \frac{n+5}{2}$$ #

Example 3   By the above conclusions, one can check that

$$g(x_1, x_2, \cdots, x_{17}) = \sum_{i=9}^{16} f_i^{17}(x_1, \cdots, x_{17})$$

To ensure that the key stream of a stream cipher of the type of Fig 3. 2 has large linear complexity, one has to use Boolean functions with high nonlinear order as combining ones. By summarizing the foregoing conclusions, we see that SML functions are ideal combining ones for this kind of stream cipher. On the other hand, SML functions can be modified into more suitable combining functions without changing their spectrum distributions. Setting

$$S_{(G)}(w) = S_{(g)}(\overline{w})$$

where $\overline{w} = (w_1 + 1, \cdots, w_n + 1)$. It follows from Theorem 4. 5a that

$$G(x_1, \cdots, x_n) = g(x_1, x_2, \cdots, x_n) + \sum_{i=1}^{n} x_i$$

Actually, $G(x_1, x_2, \cdots, x_n)$ can be expressed as

$$G(x) = \begin{cases} 1, & if \quad W_H(x) > \dfrac{n}{2} \text{ and } W_H(x) \text{ is even} \\ & or \quad W_H(x) < \dfrac{n}{2} \text{ and } W_H(x) \text{ is odd}; \\ 0, & if \quad W_H(x) > \dfrac{n}{2} \text{ and } W_H(x) \text{ is odd} \\ & or \quad W_H(x) < \dfrac{n}{2} \text{ and } W_H(x) \text{ is even} \end{cases}$$

It can also be seen that $G(x_1, x_2, \cdots, x_n)$ is correlation immune of order 1, but slightly unbalanced.

5 The Stability of Linear Complexity of Sequences

Linear complexity of the key stream of a stream cipher is an important index for measuring the strength or security of the stream cipher. But large linear complexity of the key stream does not guarentee the security of a stream cipher. To make a stream cipher secure, one has to make the linear complexity of the key stream not only large, but also stable. In order to measure the linear-complexity stability of sequences, we have already introduced the weight complexity and sphere complexity as measure indexes in Section 4. 3. The main purpose of this chapter is to develop bounds on the weight complexities of various kinds of sequences. Section 5. 1 presents basic results about linear complexity of sequences. Section 5. 2 is devoted to the stability of linear complexity of binary sequences with period 2^a, by giving many bounds on the weight complexity of these sequences. Since many kinds of key stream sequences are produced from ML-sequences by means of filtering or non-linear combining or clock-controlling, it is necessary to make clear the stability of linear complexity of ML-sequences before analyzing that of the produced key streams. Section 5. 3 develops lower bounds on the weight complexity of binary ML-sequences. Based on the lower bounds of the weight complexity of ML-sequences, Section 5. 4 gives lower bounds on the linear complexity of nonlinear filtered ML-sequences. Due to their large linear complexities, clock controlled ML-sequences occupy an important position in stream ciphers. Consequently, it is urgent to cultivate the linear-complexity stability of this kind of sequence. For this purpose, Section 5. 5 develops bounds for the weight complexity of these sequences. Owing to the merits of both clock controlled and nonlinear filtered binary ML-sequences, a new kind of key-stream generator is proposed and a lower bound on the linear complexity of the output sequences is derived from that of weight complexity of clock-controlled ML-sequences. Since the stability of linear complexity of sequences is of great importance, Section 5. 7 gives another approach to it by introducing another two measure indexes, i. e. , fixed-complexity distance (FCD) and variable-complexity distance (VCD). Furthermore, the relationships between weight complexity and fixed-complexity distance as well as sphere complexity and varible-complexity distance are established by using Blahut's theorem. Bounds on the fixed complexity distance of binary sequences with period 2^n are also developed in this section.

5. 1 Linear Complexity and Sequences

5. 1. 1 Linear Complexity and Finite Sequences

We shall write s^n to denote a sequence $s_0 s_1 \cdots s_{n-1}$ of length n whose components lie in a finite field $GF(q)$. In the case $n=0$, b^0 means the empty sequence, in the case $n=\infty$, b^∞ denotes the semi-infinite sequence $s_0 s_1 \cdots s_n \cdots$. The linear complexity of s^n with respect to $GF(q)$, denoted as $L(s^n)$, is defined as the smallest nonnegative integer L such that there exist $c_1, c_2, \cdots c_L$ in $GF(q)$ for which

$$s_j + c_1 s_{j-1} + \cdots + c_L s_{j-L} = 0$$
$$for \quad L \leqslant j < n \tag{1}$$

where we require $c_1, c_2, \cdots, c_L$ and s_i to be in the same field. This definition is nearly the same as saying that $L(s^n)$ is the order of the homogeneous linear recursion of least order satisfied by s^n; the slight difference is that we do not require c_L to be non-zero so that the linear recursion (1) might have order less than L.

It follows immediately from this definition that

$$L(s^a) = 0 \quad \text{iff} \quad b^a = 0^a \tag{2}$$

where 0^a here and hereafter denotes the all-zero sequence of length n and where "iff" stands for "if and only if". Another immediate consequence is that, for $0 < n < \infty$,

$$L(s^a) = n \quad \text{iff} \quad s^{a-1} = 0^{a-1}$$

$$\text{and } s_{a-1} \neq 0 \tag{3}$$

The engineering interpretation of $L(s^a)$ is as the length L of the shortest linear feedback shift register (LFSR) that can generate the sequence, see Fig. 5.1, such a LFSR is said to be non-singular when $c_1 \neq 0$, i. e. , when it corresponds to a linear recursion of order L.

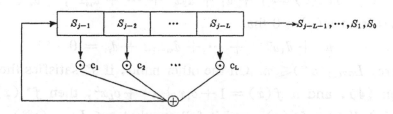

Fig. 5. 1   The linear feedback shift register interpretation
        of the recursion (1)

There is an efficient algorithm for computing $L(s^a)$ and determining the coefficients $c_1, c_2, \cdots, c_L$ in (1) with $L = L(s^a)$ [Mass 69]. This algorithm is commonly called the "Berlekamp-Massey" (B-M) algorithm. It is also very useful in decoding the BCH codes [Blah 83, Macw 77]. The mean of linear complexity of finite sequences was exploited by Rueppel [Ruep 86, pp. 36-41].

We now introduce a kind of new linear complexity. Suppose that $s^a = s_0 s_1 \cdots s_{a-1}$ is a sequence of length n whose components are lie in the field $GF(q^m)$. The linear complexity of s^a with respect to the subfield $GF(q)$, here and hereafter denoted as $L_{GF(q)}(s^a)$, is defined as the smallest nonnegative integer L such that there exists $c_1, c_2, \cdots, c_L$ in GF

(q) for which

$$s_j + c_1 s_{j-1} + \cdots + c_L s_{j-L} = 0$$
$$\text{for } L \leqslant j < n \tag{4}$$

In this definition, we require the coefficients $c_1, c_2, \cdots, c_L$ to be in a smaller field $GF(q)$. This is the difference between $L(s^n)$ and $L_{GF(q)}(s^n)$. Obviously, the following inequality holds:

$$L(s^n) \leqslant L_{GF(q)}(s^n) \tag{5}$$

For instance, let u be a primitive element of $GF(2^m)$, as

$$s^\infty = u^0 u^1 u^2 u^3 \cdots .$$

By definition, it is apparent that $L(s^\infty) = 1$. Suppose that the minimal polynomial of u over $GF(2)$ is $m(x)$, and the reciprocal polynomial $m^*(x)$ of $m(x)$ is

$$m^*(x) = 1 + d_1 + d_2 x^2 + \cdots + d_m x^m, \quad d_i \in GF(2)$$

It follows from $m(u) = 0$ that

$$u^m + d_1 u^{m-1} + \cdots + d_{m-1} u + d_m = 0 \tag{6}$$

Therefore, $L_{GF(2)}(s^\infty) \leqslant m$. On the other hand, if s^∞ satisfies the linear recursion (4), and if $f(x) = 1 + c_1 x + \cdots + c_L x^L$, then $f^*(x) = 0$. Thus $m(x)$ divides $f^*(x)$, and it follows that $m \leqslant L_{GF(2)}(s^\infty)$. Hence $L_{GF(2)}(s^\infty) = m$. This example shows the difference clearly.

It is well known that $GF(q^m)$ can be regarded as a linear space of dimension m over $GF(q)$. Let $u_1, u_2, \cdots, u_m$ be a basis for $GF(q^m)$, then each element u of $GF(q^m)$ can be expressed as

$$u = \sum_{i=1}^m a_i u_i, \quad a_i \in GF(q)$$

Assume that for every j

$$s_j = \sum_{i=1}^m s_{i,j} u_i \quad s_{i,j} \in GF(q) \tag{7}$$

then the recursion (4) is equivalent to

$$s_{i,j} + c_1 s_{i,j-1} + \cdots + c_L s_{i,j-L} = 0$$
$$\text{for } L \leqslant j \leqslant n \qquad i = 1, 2, \cdots, m \tag{8}$$

This demonstrates that the linear complexity $L_{GF(q)}(s^n)$ with respect to $GF(q)$ is the least length of the LFSRs which can generate the lower field sequences $s_{i,0} s_{i,1} \cdots s_{i,n-1}, i = 1, 2, \cdots, m$, at the same time only with different initial states. It is called the LFSR synthesis of multisequences. J. L. Massey proposed a conjectured algorithm for the LFSR synthesis of multisequences. The algorithm was proved to be a universal one for the minimal realization of any linear system by Ding [Ding 87a] (see Appendix A). We would like to point out that Massey' s algorithm can be used to determine $L_{GF(q)}(s^n)$ and the coefficients $c_1, c_2, \cdots, c_L$ in (4).

The dynamic behaviour of $L(s^n)$ for binary sequences as n increases was exploited by Rueppel [Ruep 86]. As for the stability of linear complexity of finite sequences and of lower field linear complexity of finite sequences, it is still remained to be investigated.

5. 1. 2 Linear Complexity and Periodic Sequences

From the viewpoint of cryptography, our main interest will be in the stability of linear complexity of periodic sequences. A semi-infinite sequence s^∞ is said to be periodic with period N if there is a positive integer N such that

$$s_j = s_{j+N} \text{ for all } j \geqslant 0 \tag{9}$$

The smallest N that satisfies (7) is called the minimal period of s^∞. A periodic sequence s^∞ of period N is completely characterized by its first period $s^N = s_0 s_1 \cdots s_{N-1}$.

It is well known that if s^∞ is periodic with period N, then $L(s^\infty) \leqslant N$ and there are unique coefficients $c_1, c_2, \cdots, c_L$ for which (1) is satisfied for all $j \geqslant 1$ with $L = L(s^\infty)$, moreover, $c_L \neq 0$. It follows that the linear complexity of a periodic sequence is precisely the order of the homogeneous linear recursion (1) of least order satisfied by the sequence.

Let s^∞ be a sequence of period N over $GF(q)$. The polynomial

$$f(x) = c_0 + c_1 x + \cdots + c_n x^n,$$

$$c_0 = 1, \quad c_n \neq 0, \quad c_i \in GF(q)$$

is called the minimal polynomial of s^∞ if the following linear recurrence relation

$$s_v = c_1 s_{v-1} + c_2 s_{v-2} + \cdots$$
$$+ c_n s_{v-n}, \quad v \geqslant n$$

is satisfied by s^∞ and $deg(f)$ is minimal. Denoting here and hereafter f_s as the minimal polynomial of s^∞, then $deg(f_s)$ is the linear complexity (linear equivalence) of s^∞, i. e. , the length of the shortest LFSR that can generate s^∞.

The formal power series or the generating function over $GF(q)$ of the sequence s^∞ is defined as

$$s(x) = \sum_{i=0}^{\infty} s_i x^i$$

If s^∞ is periodic with period N, then

$$(1 - x^N)s(x) = s^N(x) \qquad (10)$$

where $s^N(x) = s_0 + s_1 x + \cdots + s_{N-1} x^{N-1}$. Hence, the following result holds:

Theorem 5. 1 The generating function of every periodic sequence can be expressed as

$$s(x) = g(x)/f(x)$$

with $f(0) \neq 0$ and $deg(g) < deg(f)$

Noting that

$$f_s(x)s(x) = \sum_{k=0}^{\infty} \left(\sum_{0 \leqslant i \leqslant min\{n,k\}} c_i s_{k-i} \right) x^k$$

and

$$\sum_{0 \leqslant i \leqslant min\{n,k\}} c_i s_{k-i} = 0 \quad for \ all \ k \geqslant n$$

where $f_s(x) = c_0 + c_1 x + \cdots + c_n x^n$ is the minimal polynomial of s^∞, we have

$$f_s(x)s(x) = \sum_{k=0}^{n-1} \left(\sum_{0 \leqslant i \leqslant k} c_i s_{k-i} \right) x^k$$
$$\overset{\triangle}{=} r_s(x) \qquad (11)$$

and we conclude that the greast common divisor (gcd) of f_s and r_s is 1. If not so, setting

$$g(x) = f_s(x)/gcd(f_s, r_s)$$
$$= g_0 + g_1 x + \cdots + g_m x^m$$

and

$$h(x) = r_s(x)/gcd(f_s, r_s)$$

implies that $deg(h) < deg(g)$ and $g(x)s(x) = h(x)$. This means that

$$\sum_{0 \leqslant i \leqslant m} g_i s_{k-i} = 0 \qquad \text{for all} \qquad k \geqslant m$$

This is contrary to the minimality of f_s. Thus, $gcd(f_s, r_s) = 1$.

Conversely, assume that $s(x) = g(x)/f(x)$ is a rational form of $s(x)$ with $gcd(f, g) = 1$ and $f(0) = 1$; we now prove that $f(x)$ is the minimal polynomial of s^∞. It is apparent that $(f(x), deg(f))$ is a LF-SR that can generate the sequence s^∞. Therefore the minimal polynomial f_s divides f. Let $f = f_s \cdot h$, then we have $g(x) = f(x)s(x) = hf_s s(x) = h(x)r_s(x)$. Since $gcd(f, g) = 1$, so $h(x)$ must be equal to 1. Hence, $f = f_s$ is the minimal polynomial of s^∞. Thus, we have proved the following theorem:

Theorem 5. 2   Let

$$s(x) = r(x)/f(x), \quad deg(f) > deg(r), \quad f(0) = 1$$

a rational form of $s(x)$. Then $f(x)$ is the minimal polynomial of s^∞ iff $gcd(f(x), r(x)) = 1$.   #

Theorem 5. 3   Let s^∞ be periodic with period N. Then the minimal polynomial of s^∞ is given by

$$f_s(x) = (1 - x^N)/gcd(s^N(x), 1 - x^N)$$

Proof: It is obvious that $s^N(x)/(1-x^N)$ is a rational form of $s(x)$, so is $g(x)/f(x)$. Where $g(x) = s^N(x)/gcd(s^N(x), 1-x^N)$ and $f(x) = (1-x^N)/gcd(s^N(x), 1-x^N)$.

It follows, from Theorem 5. 3 and $gcd(g, f) = 1$, that $g(x) = (1 - x^N)/gcd(1 - x^N, s^N(x))$ is the minimal polynomial of s^∞.   #

Definition 5.4 A rational form of $s(x) = g(x)/f(x)$ is said to be reduced if $gcd(f,g) = 1$.

Theorem 5.5 Let the reduced rational forms of s^∞ and t^∞ be respectively

$$s(x) = r_s(x)/f_s(x)$$
$$t(x) = r_t(x)/f_t(x)$$

then

$$f_{s+t} = f_s f_t / gcd(f_s f_t, \ f_s r_t + r_s f_t)$$

Proof: Noticing that

$$s(x) + t(x) = (f_s r_t + r_s f_t)/f_s f_t$$

and $gcd(f_s f_t, \ gcd(f_s f_t, \ f_s r_t + r_s f_t)) = 1$, it follows from Theorem 5.2 that Theorem 5.5 holds.   #

5.2 Weight Complexity and Lower Bounds for the Weight Complexity of Binary Sequences With Period 2^n

Before developing lower bounds for the weight complexities of binary sequences with period 2^n, we would like to give a general result of weight complexity and sphere complexity of sequence over $GF(q)$.

Theorem 5.6 Let s^∞ and w^∞ be periodic with period N over $GF(q)$, then

$$WC_u(s^\infty) = \min_{W_H(w^N)=u} deg \ \frac{f_s f_w}{gcd(f_s f_w, f_s r_w + r_s f_w)} \tag{12}$$

$$SC_u(s^\infty) = \min_{0<W_H(w^N)\leqslant u} deg \ \frac{f_s f_w}{gcd(f_s f_w, f_s r_w + r_s f_w)} \tag{13}$$

where here and hereafter $r_s(x) = f_s(x)s(x)$ denotes the numerator polynomial.

Especially, we have

$$WC_1(s^\infty) = \min_{\substack{0 \leqslant i \leqslant N-1 \\ a \in GF(q)-\{0\}}} deg \frac{(1 - x^N)}{gcd(1 - x^N, ax^i + r_s(x)g(x))} \quad (14)$$

where $g(x) = (1 - x^N)/f_s(x)$.

Proof: It is trival from Definition 4. 10 and Theorem 5. 5 that (12) and (13) in Theorem 5. 6 hold. To prove (14), let w^∞ be periodic with period N, and $w^N(x) = ax^i$, $0 \leqslant i \leqslant N-1$, then it is obvious that $f_w = 1 - x^N$, and $r_w = ax^i$. Therefore, it follows from (12) that (14) holds. #

5. 2. 1 Weight Complexity $WC_1(s^\infty)$ and Lower Bounds on $WC_1(s^\infty)$ of Binary Sequences with Period 2^n

First of all, let us stipulate that the sequences discussed in Section 5. 2. 1 are binary and the addition "$+$" denotes modulo-2 addition.

Theorem 5. 7   Let s^∞ be a sequence of period 2^n . If $L(s^\infty) < 2^n$, then
$$WC_1(s^\infty) = 2^n$$

Proof: It follows from Theorem 5. 6 that
$$WC_1(s^\infty) = \min_{0 \leqslant i \leqslant 2^n-1} deg \frac{(1 + x)^{2^n}}{gcd((1 + x)^{2^n}, x^i + r_s(x)(1 + x)^{2^n - L(s^\infty)})}$$

If $L(s^\infty) < 2^n$, then for every $0 \leqslant i \leqslant 2^n - 1$
$$gcd[(1 + x)^{2^n}, x^i + r_s(x)(1 + x)^{2^n - L(s^\infty)}] = 1$$

Hence $WC_1(s^\infty) = 2^n$. #

Notice there are $2^{2^n} - 2^{2^{n-1}}$ binary sequences of period 2^n with linear complexity less than 2^n. For any such a sequence, one bit of change in the corresponding places of every period will make the linear complexity of the sequence jump to 2^n. This also shows that there are a lot of sequences of period 2^n with linear complexity 2^n whose linear complexity stability is very bad. For example, let $s^{N-1} = 0^{N-1}$, and $s_{N-1} = 1$, where $N = 2^n$. Then $L(s^\infty) = 2^n$. It is apparent that $WC_1(s^\infty) = 0$.

Lemma 5.8 Let $k = 2^L k_1$ with k_1 odd and L a positive integer, then $(1 + x)^{2^L} \| x^k + 1$, i.e., $(1 + x)^{2^L}$ divides $x^k + 1$, but $(1+x)^{2^L+1}$ does not.

Proof: Notice

$$x^k + 1 = x^{2^L k_1} + 1$$
$$= (1 + x)^{2^L}(x^{2^L(k_1 - 1)} + \cdots + x^{2^L} + 1)$$
$$= (1 + x)^{2^L} u(x)$$

and $u(1) = k_1 (mod\ 2) = 1$, whence the conclusion of Lemma 5.8 is true.   #

Theorem 5.9 Let s^∞ be periodic with period 2^n. If $L(s^\infty) = 2^n$ and $r_s = x^j$, $0 \leqslant j \leqslant 2^n - 1$. Then $WC_1(s^\infty) = 0$ and

$$\min_{\substack{W_H(w^N)=1 \\ w^N \neq s^N}} L(s^\infty + w^\infty) = 2^{n-1}$$

Proof: Setting $i = j$, then

$$gcd[(1 + x)^{2^n}, x^i + r_s(x)(1 + x)^{2^n - L(s^\infty)}] = (1 + x)^{2^n}$$

Thus, it follows from Theorem 5.6 that $WC_1(s^\infty) = 0$.

For any k with $1 \leqslant k \leqslant 2^n - 1$, let $k = 2^l k_1$ with k_1 odd. It follows from Lemma 5.9 that $x^k + 1$ has root 1 of order 2^e, but the maximum value for e is $n - 1$. Whence $x^k + 1$ has root 1 of order at most 2^{n-1}. If $j < 2^{n-1}$, then choose $i = 2^{n-1} + j \leqslant 2^n - 1$, Therefore $x^i + x^j = x^j(1+x)^{2^{n-1}}$; if $j = 2^{n-1}$, choose $i = 0$, then $x^i + x^j = (1 + x)^{2^{n-1}}$; if $j > 2^{n-1}$, choose $i = j - 2^{n-1} > 0$, Then $x^i + x^j = x^i(1 + x^{j-i}) = x^i(1 + x)^{2^{n-1}}$. Hence, we obtain

$$\max_{i \neq j} deg\ gcd[((1 + x)^{2^n}, x^i + r_s(x + 1)^{2^n - L(s^\infty)}]$$
$$= gcd((1 + x)^{2^n}, x^i + x^j) = 2^{n-1}$$

It follows from this fact and Theorem 5.6 that Theorem 5.9 holds.

The following lemma is obviously true. Although simple, it will play a key role in the proofs of most of the theorems in this section.

Lemma 5.10 If $(1+x)^{k_1} \| h_1(x)$, $(1+x)^{k_2} \| h_2(x)$ and $k_1 \neq k_2$, then

$$(1 + x)^{\min\{k_1, k_2\}} \| h_1(x) + h_2(x)$$

Theorem 5.11 Let s^∞ be periodic with period 2^n. If $L(s^\infty) = 2^n$ and

$$r_s = 1 + x^j + x^{2^{n-1}}$$

where $j = 2^e j_1$ with j_1 odd; $0 < j \leqslant 2^n - 1$ and $j \neq 2^{n-1}$, then

$$WC_1(s^\infty) = 2^{n-1} - 2^e$$

Proof: Let $|i-j| = 2^d t$ with t odd. Noticing $j = 2^e j_1$, $h(x) = r_s(x) + x^i$ can be written as

$$h(x) = \begin{cases} (1+x)^{2^{n-1}}, & \text{if } i = j \\ (1+x)^{2^d}[(1+x)^{2^{n-1}-2^d} + x^j u(x)], & \text{if } i > j \\ (1+x)^{2^d}[(1+x)^{2^{n-1}-2^d} + x^i u(x)], & \text{if } i < j \end{cases}$$

where $u(x) = 1 + x^{2^d} + \cdots + x^{(t-1)2^d}$. To ensure that $h(x)$ has root 1 of maximum order, it follows from Lemma 5.10 that d must be equal to $n-1$. Consequently, t must be equal to 1. Whence $u(x) = 1$ and $i = 2^{n-1} + j = 2^e(2^{n-1-e} + j_1) = 2^e i_1$. Since $j \neq 2^{n-1}$, i_1 is odd. Thus, we get

$$h(x) = \begin{cases} (1+x)^{2^{n-1}}, & \text{if } i = j \\ (1+x)^{2^{n-1}+2^e} u_1(x), & \text{if } i > j \\ (1+x)^{2^{n-1}+2^e} u_2(x), & \text{if } i < j \end{cases}$$

where $u_1(x)$ and $u_2(x)$ are polynomials such that $u_1(1) = u_2(1) = 1$.

On the other hand, if $0 < j < 2^{n-1}$, then choose $i = 2^{n-1} + j \leqslant 2^n - 1$; if $j > 2^{n-1}$, choose $i = j - 2^{n-1}$. In both cases we can make $h(x)$ have root 1 with maximum order $2^{n-1} + 2^e$. Whence it follows from Theorem 5.6 that

$$\begin{aligned} WC_1(s^\infty) &= 2^n - max \ deg \ gcd[(1+x)^{2^e}, h(x)] \\ &= 2^n - 2^{n-1} - 2^e \\ &= 2^{n-1} - 2^e \end{aligned}$$

\#

Theorem 5.12 Let s^∞ be periodic with period 2^n. If $L(s^\infty) = 2^n$ and

$$r_s(x) = 1 + x^{2^{t_1}} + \cdots + x^{2^{t_v}}, \qquad 4 \leqslant v \leqslant n - 1$$

Where v is even, and $t_1 < t_2 < \cdots < t_v$, then

$$WC_1(s^\infty) = 2^n - 2^{t_1} - 2^{t_2}$$

Proof: Let $h(x) = r_s(x) + x^i$, then we have

$$\begin{aligned} h(x) &= (1+x)^{2^{t_1}} + \cdots + (1+x)^{2^{t_v}} + x^i + 1 \\ &= (1+x)^{2^{t_1}}[1 + (1+x)^{2^{t_2}-2^{t_1}} + \cdots \end{aligned}$$

$$+ (1 + x)^{2^{t_v} - 2^{t_1}}] + x^i + 1$$

In order to make $h(x)$ have root 1 of maximum order, it follows from Lemma 5. 10 that we must choose $i = 2^{t_1}e$. Consequently, we obtain

$$h(x) = (1 + x)^{2^{t_1}}[(1 + x)^{2^{t_2} - 2^{t_1}} + \cdots + (1 + x)^{2^{t_v} - 2^{t_1}} + u(x)]$$

where

$$u(x) = x^{(e-1)2^{t_1}} + \cdots + x^{2^{t_1}}$$

Suppose $e - 1 = 2^k e_1$, e_1 odd, then

$$u(x) = x^{2^{t_1}}(1 + x^{2^{t_1}})(1 + x^{2 \cdot 2^{t_1}}) \cdots (1 + x^{2^{k-1} 2^{t_1}})u_1(x)$$

where

$$u_1(x) = 1 + x^{2^{t_1} + k} + \cdots + x^{(e_1 - 1) \cdot 2^{t_1} + k}$$
$$u_1(1) = 1$$

In order to make $h(x)$ have root 1 of maximum order, it follows from Lemma 5. 10 that $k = t_2 - t_1$. Consequently, we have

$$h(x) = (1 + x)^{2^{t_2}}[1 + (1 + x)^{2^{t_3} - 2^{t_2}} +$$
$$\cdots + (1 + x)^{2^{t_v} - 2^{t_2}} + x^{2^{t_1}}u_1(x)]$$

Let $e_1 - 1 = 2^p q$, q odd, then

$$V(x) = 1 + x^{2^{t_1}}u_1(x)$$
$$= (1 + x)^{2^{t_1}} + x^{2^{t_1} + 2^{t_2}}(1 + x)^{(2^p - 1)2^{t_2}}v_1(x)$$

where $v_1(x)$ is a polynomial such that $v_1(1) = 1$. Notice $(2^p - 1)2^{t_2} \neq 2^{t_1}$ for any $p \geqslant 0$, so $V(x)$ has root 1 of order at most $min\{2^{t_1}, (2^p - 1)2^{t_2}\} \leqslant 2^{t_1}$. Actually, if we choose $p = 1$ and $q = 1$, then

$$V(x) = (1 + x)^{2^{t_1}}v_2(x), \qquad v_2(1) = 1$$

Since $i = 2^{t_1}e = 2^{t_1}(2^{k+1} + 2^k + 1) \leqslant 2^n - 1$. It follows that $WC_1(s^\infty) = 2^n - 2^{t_1} - 2^{t_2}$. #

Lemma 5. 13 Assume that $s - 1 = 2^k p$, p odd, then

$$1 + y + y^2 + \cdots + y^{s-2} = (1 + y)^{2^k - 1}h_1(x)$$

where $h_1(x) = 1 + y^{2^k} + \cdots + y^{(p-1)2^k}$, and $h_1(x) = 1$.

Proof: It is not difficult to see that $\left\{\begin{pmatrix} 2^k - 1 \\ i \end{pmatrix}\right\}mod\ 2 = 1$ for every $0 \leqslant i \leqslant 2^k - 1$. Therefore,

$$(1+y)^{2^t-1} = 1 + y + y^2 + \cdots + y^{2^t-1}$$

and

$$(1+y)^{2^t-1}h_1(x) = 1 + y + y^2 + \cdots y^{2^t-1}$$
$$+ y^{2^t} + y^{2^t+1} + \cdots + y^{2^{t+1}-1}$$
$$+ \cdots$$
$$+ y^{(p-1)2^t} + \cdots + y^{p2^t-1}$$
$$= 1 + y + y^2 + \cdots + y^{s-2} \qquad \#$$

Theorem 5. 14 Let s^∞ be periodic with period 2^n and

$$r_s = 1 + x^{i_1} + \cdots + x^{i_m}$$

where m is even and $i_v = 2^{e_v} \cdot p_v$, p_v odd, $1 \leqslant v \leqslant m$ and $e_1 < e_2 < \cdots < e_m$.
Then

$$WC_1(s^\infty) \geqslant 2^n - 2^{e_1} - 2^{e_2}$$

Proof: Since m is even, so

$$h(x) = r_s(x) + x^i$$
$$= (1 + x^{i_1}) + \cdots + (1 + x^{i_m}) + (1 + x^i)$$
$$= h_1(x) + (1 + x^i)$$

Apparently, $h_1(x)$ has root 1 of order 2^{e_1}. In order to make $h(x)$ have
root 1 of maximum order, it follows from Lemma 5. 10 that $i = 2^{e_1}q$.

a) Suppose $q > p_1$, let $q - p_1 = 2^k s$, s odd and $u(x) = (1 + x^{i_1})$
$+ (1 + x^i)$, then we get from Lemma 5. 13 that

$$u(x) = (1 + x)^{2^{e_1}} x^{p_1 2^{e_1}} [1 + x^{2^{e_1}} + \cdots + x^{2^{e_1}(q-p_1-1)}]$$
$$= (1 + x)^{2^{e_1}+(2^k-1)2^{e_1}} x^{p_1 2^{e_1}} [1 + x^{2^{e_1}+k} + \cdots x^{2^{e_1}+k(s-1)}]$$

To make $h(x)$ have root 1 of maximum order, it follows from Lemma
5. 10 that $k = e_2 - e_1$. As a result,

$$u(x) + (1 + x^{i_2}) = (1 + x)^{2^{e_2}} [x^{p_1 2^{e_1}}(1 + x^{2^{e_2}} + \cdots +$$
$$x^{(s-1)2^{e_2}}) + (1 + x^{2^{e_2}} + \cdots x^{(p_2-1)2^{e_2}})]$$

a. 1) If $s = p_2$, then

$$u(x) + (1 + x^{i_2}) = (1 + x)^{2^{e_1}+2^{e_2}} d(x)$$

where $d(1) = 1$.

a. 2) If $s > p_2$, let $s - p_2 = 2^t c$, c odd, then we get from Lemma
5. 13 that

$$u(x) + (1 + x^i{}_2) = (1 + x)^{2^{e_2}}\big[(1 + x^{p_1 2^{e_1}})w_1(x)$$
$$+ x^{p_2 2^{e_2}}(1 + x)^{2^{e_2}(2^t - 1)}w_2(x)\big]$$

where $w_1(1) = w_2(1) = 1$. Notice that s and p_2 both are odd, and $s > p_2$, therefore $2^{e_2}(2^t - 1) > 2^{e_1}$. Because $2^{e_3} > 2^{e_1} + 2^{e_2}$, $u(x) + (1 + x^i{}_2)$ has root 1 of order at most $2^{e_1} + 2^{e_2}$.

a. 3) If $s < p_2$, let $p_2 - s = 2^t c$, c odd, then we get

$$u(x) + (1 + x^i{}_2) = (1 + x)^{2^{e_2}}\big[(1 + x)^{p_1 2^{e_1}}(1 + x^{2^{e_2}} + \cdots$$
$$+ x^{(s-1)2^{e_2}}) + x^{s2^{e_2}}(1 + x^{2^{e_2}} + \cdots$$
$$+ x^{(p_2-s-1)2^{e_2}})\big]$$
$$= (1 + x)^{2^{e_2}}\big[(1 + x^{p_1 2^{e_1}})w_1(x) + x^{s2^{e_2}}w_2(x)\big]$$

Since s is odd, $w_1(x) = 1$. For the same reason as in case a. 2) we know that $h(x)$ has root 1 of order at most $2^{e_1} + 2^{e_2}$.

We have so far proved that in case a) $h(x)$ has root 1 of order at most $2^{e_1} + 2^{e_2}$.

b) Suppose $q < p_1$, let $p_1 - q = 2^k s$, s odd, similarly, we can prove that $h(x)$ has root 1 of order at most $2^{e_1} + 2^{e_2}$.

c) Suppose $q = p_1$, then $h(x)$ has root 1 of order 2^{e_2}.

Summarizing cases a), b) and c), we see that the order of root 1 of $h(x)$ is less than or equal to $2^{e_1} + 2^{e_2}$. Hence, it follows from Theorem 5.6 that $WC_1(s^\infty) \geqslant 2^n - 2^{e_1} - 2^{e_2}$ #

In the foregoing Theorem 5.12 and 5.14, we assume $e_i \neq e_j$ $(i \neq j)$, and prove that the linear complexity does not decrease too much after changing any bit in every period in the corresponding places. We now consider the case $e_i = e_j$, and prove that the linear complexity will decrease much faster than that in the case $e_i \neq e_j$.

Theorem 5.15 Let s^∞ be periodic with period 2^n. If $L(s^\infty) = 2^n$ and $r_s(x) = 1 + x^{j_1} + x^{j_2}$ with $j_1 = 2^e t_1$, $j_2 = 2^e t_2$, $j_1 < j_2$, $t_2 - t_1 = 2^k m$, where t_1, t_2 and m are odd. Then

$$WC_1(s^\infty) = 2^n - 2^e(2^k + 1)$$

Proof: Notice that $j_2 > j_1$, so $t_2 > t_1$. Because t_1 and t_2 both are odd, therefore $k \geqslant 1$. It follows from Lemma 5.13 that

$$h(x) = (1+x)^{2^{e+k}} x^{i_1 2^e} \cdot$$
$$(1 + x^{2^{e+k}} + \cdots + x^{(m-1)2^{e+k}}) + 1 + x^i$$

To make $h(x)$ has root 1 with maximum order, by Lemma 5.10 we choose $i = 2^{e+k} i_1$, i_1 odd.

a) If $i_1 = m$, then we get

$$h(x) = (1+x)^{2^e(2^k+1)} h_1(x)$$

where

$$h_1(x) = (1 + x^{2^e} + \cdots + x^{(i_1-1)2^e})$$
$$\cdot (1 + x^{2^{e+k}} + \cdots + x^{(m-1)2^{e+k}})$$
$$h_1(1) = 1$$

This demonstrates that $h(x)$ has root 1 of order $2^e(2^k + 1)$.

b) If $i_1 > m$, let $i_1 - m = 2^p q$, q odd. Notice i_1 and m both are odd, and $i_1 > m$, whence $p \geqslant 1$. By Lemma 5.13 we obtain

$$h(x) = (1+x)^{2^{e+k}}[(1 + x^{2^e i_1})(1 + x^{2^{e+k}} + \cdots + x^{(m-1)2^{e+k}})$$
$$+ x^{m \cdot 2^{e+k}}(1 + x^{e+k} + \cdots + x^{(i_1-m-1)2^{e+k}})]$$
$$= (1+x)^{2^{e+k}}[(1 + x^{2^e i_1})u_1(x) + x^{m2^{e+k}}(1 + x)^{2^{e+k}(2^p-1)} v_1(x)]$$

where $u_1(1) = v_1(1) = 1$. Since $p \geqslant 1$, so $2^{e+k}(2^p - 1) > 2^e$, it follows from Lemma 5.10 that

$$(1 + x^{2^e i_1})u_1(x) + x^{m2^{e+k}}(1 + x)^{2^{e+k}(2^p-1)} v_1(x)$$

has root 1 of order 2^e. Whence $h(x)$ has root 1 of order $2^e(2^k + 1)$.

c) If $i_1 < m$, similarly, we can prove $h(x)$ has root 1 of order $2^e(2^k + 1)$.

On the other hand, it is possible that we can choose $i_1 = m$, then $j_2 - j_1 = i$. Hence

$$WC_1(s^\infty) = 2^n - 2^e(2^k - 1) \qquad\qquad \#$$

5.2.2 Weight Complexity $WC_2(s^\infty)$ and Lower Bounds on $WC_2(s^\infty)$

In Section 5.2.1 we have discussed the weight complexity $WC_1(s^\infty)$ for binary sequences with period 2^n. Now we shall investigate the weight complexity $WC_2(s^\infty)$ for binary sequences with period 2^n. Let us still stipulate that all the sequences discussed in Section 5.2.2 and 5.2.3 are binary and all the operations are in $GF(2)$, i.e., "$+$"denotes modulo-2 addition.

Theorem 5.16 Let s^∞ be periodic with period 2^n, then
$$WC_2(s^\infty) =$$

$$\min_{i+2^k p \leqslant 2^n-1} \deg \frac{f_s(x)(1+x)^{2^n-2^k}}{gcd(f_s(x) \cdot (1+x)^{2^n-2^k}, x^i g(x) f_s(x) + r_s(x)(1+x)^{2^n-2^k})}$$

where
$$g(x) = x^{2^k(p-1)} + \cdots + x^{2^k} + 1$$

Proof: Let u^∞ and v^∞ be periodic with period 2^n, and $W_H(u^N) = W_H(v^N) = 1$, $u_i = 1$ and $v_j = 1 (i \neq j)$. Set $w^\infty = u^\infty + v^\infty$. Notice that $f_u(x) = (1+x)^{2^n}$, $r_u(x) = x^i$ and $f_v(x) = (1+x)^{2^n}$, $r_v(x) = x^j$. We get
$$W(x) = (x^i + x^j)/(1+x)^{2^n}$$
Whence, we obtain from Theorem 5.3 that
$$f_w(x) = (1+x)^{2^n}/gcd[(1+x)^{2^n}, x^i + x^j]$$
Without loss of generality, we assume that $i < j$, and $j-i=2^k p$, with p odd. Then we have
$$x^i + x^j = x^i(1+x)^{2^k}(x^{2^k(p-1)} + \cdots + x^{2^k} + 1)$$
Therefore
$$f_w(x) = (1+x)^{2^n-2^k}$$
$$r_w(x) = x^i(x^{2^k(p-1)} + \cdots + x^{2^k} + 1)$$
Thus, by Theorem 5.6 we see that the conclusion of Theorem 5.16 is valid. #

Theorem 5.17 Let S^∞ be periodic with period 2^n. If $L(S^\infty)=2^n$, then $WC_2(S^\infty)=2^n$.

Proof: Since $L(S^\infty)=2^n$, we have
$$maxgcd\left[(1+x)^{2^n}f_s(x), (1+x)^{2^n}f_s(x)g(x) + (1+x)^{2^n}r_s(x)\right]$$

$$= (1+x)^{2^a}$$

It follows from Theorem 5.16 that

$$WC_2(S^\infty) = deg[f_s(x)(1+x)^{2^a}/maxgcd[(1+x)^{2^a}f_s(x),$$
$$f_s(x)g(x)(1+x)^{2^a} + (1+x)^{2^a}f_s(x)]]$$
$$= 2^a \qquad\qquad \#$$

Theorem 5.18 Let S^∞ be periodic with period 2^a.

a) If $L(S^\infty) < 2^{a-1}$, then $WC_2(S^\infty) = 2^{a-1}$;

b) If $L(S^\infty) > 2^{a-1}$, and $L(S^\infty) \neq 2^a - 2^m$ for every m with $0 \leqslant m \leqslant n-1$. Then $WC_2(S^\infty) = L(S^\infty)$.

Proof: a) Since $L(S^\infty) < 2^{a-1}$, so $L(S^\infty) + 2^k < 2^a$ for $0 \leqslant k \leqslant n-1$. Therefore, we get from Lemma 5.10 that

$$maxgcd[(1+x)^{2^a}f_s(x),(1+x)^{2^a}f_s(x)g(x) + (1+x)^{2^a}r_s(x)]$$
$$= (1+x)^{2^{a-1}}f_s(x)$$

Thus, we get from Theorem 5.16 that

$$WC_2(S^\infty) = (2^a + L(S^\infty)) - (2^{a-1} + L(S^\infty))$$
$$= 2^{a-1}$$

b) Notice $2^a - 2^m \neq L(S^\infty)$ for every m with $0 \leqslant m \leqslant n-1$, and $L(S^\infty) + 2^{a-1} > 2^a$. Therefore we obtain

$$maxgcd[(1+x)^{2^a}f_s(x),(1+x)^{2^a}f_s(x)g(x) + r_s(x)(1+x)^{2^a}]$$
$$= (1+x)^{2^a}$$

Hence, by Theorem 5.16 we have

$$WC_2(S^\infty) = L(S^\infty) + 2^a - 2^a = L(S^\infty) \qquad \#$$

Theorem 5.19 Let S^∞ be a sequence of period 2^a. If $L(S^\infty) = 2^{a-1}$, then the minimal period of S^∞ is 2^{a-1}. Let $T^\infty = S^\infty$ and regard T^∞ as a sequence of period 2^{a-1}. Then

$$WC_2(S^\infty) = WC_1(T^\infty) + 2^{a-1}$$

Proof: To make

$$gcd[(1+x)^{2^a}f_s(x),f_s(x)g(x)(1+x)^{2^a} + (1+x)^{2^a}r_s(x)]$$

have root 1 with maximum order, it follows from Lemma 5.10 that $k = n - 1$. Since $i + 2^k p < 2^a$, p must be equal to 1. Whence, we get from Lemma 5.8 that

$$WC_2(S^\infty) = \min_{0 \leqslant i \leqslant 2^{n-1}-1} deg[(1+x)^{2^i}/gcd[(1+x)^{2^{n-1}}, x^i + r_s(x)]]$$

$$= 2^{n-1} + \min_{0 \leqslant i \leqslant 2^{n-1}-1} deg[(1+x)^{2^i}/gcd[(1+x)^{2^{n-1}},$$

$$x^i + r_s(x)]]$$

$$= 2^{n-1} + WC_1(T^\infty) \qquad\qquad \#$$

Remark: When defining the weight complexity $WC_k(S^\infty)$ and sphere complexity $SC_k(S^\infty)$, we did not require the period of the sequence regarded to be the minimal period. In Section 5. 2. 1, lower bounds for $WC_1(t^\infty)$ have already been given. By employing Theorem 5. 19, weight complexity $WC_2(s^\infty)$ and lower bounds on it for the sequences with $L(S^\infty) = 2^{n-1}$ when regarding them as sequences of period 2^n can be given. To save space, we do not write them out here.

Theorem 5. 20 Let S^∞ be periodic with period 2^n. If $L(S^\infty) = 2^n - 2^m$, $0 \leqslant m \leqslant n-1$, and $1 + r_s(x)$, denoted as $V(x)$, has root 1 of order R with $R \neq 2^m(2^c - 1)$ for each c, $0 \leqslant c \leqslant n-m-1$, then $WC_2(S^\infty) \geqslant 2^{n-1}$.

Proof: Since $L(S^\infty) = 2^n - 2^m$, we have

$$u(x) = maxgcd[(1+x)^{2^i}f_s(x), f_s(x)g(x)(1+x)^{2^i}$$

$$+ (1+x)^{2^i}r_s(x)]$$

$$= maxgcd[(1+x)^{2^{i+1}-2^m}, g(x)(1+x)^{2^i-2^m-2^i}$$

$$+ (1+x)^{2^i}r_s(x)]$$

To make $u(x)$ have root 1 with maximum order, it follows from Lemma 5. 10 that $k = m$. Then we obtain

$$u(x) = (1+x)^{2^i}maxgcd[(1+x)^{2^i-2^m}, g(x) + r_s(x)]$$

where $g(x) = 1 + x^{2^n} + \cdots + x^{2^n(p-1)}$. For the purpose of making $u(x)$ have root 1 with maximum order, p must be odd. Let $p - 1 = 2^c d$ with d odd. Since $i + 2^m p \leqslant 2^n - 1$, so $0 \leqslant p \leqslant 2^{n-m} - 1$ and $c < n - m - 1$. By Lemma 5. 13 we get

$$g(x) - 1 = x^{2^n}(1+x)^{2^n(2^c-1)}g_1(x)$$

where $g_1(1) = 1$. Because $R \neq 2^m(2^c - 1)$ for $0 \leqslant c \leqslant n-m-1$, it follows from Lemma 5. 10 that $g(x) + r_s(x)$ has root 1 with order $min\{2^m(2^{n-m-1} - 1), R\}$. Noticing $R \leqslant 2^n - 2^m - 1$, it follows from

Theorem 5. 16 that

$$WC_2(S^\infty) = 2^n - 2^m - min\{2^{n-1} - 2^m, R\}$$
$$\geqslant 2^{n-1} \qquad\qquad \#$$

5. 2. 3 Weight Complexity $WC_n(S^\infty)$ and Lower Bounds for $WC_n(S^\infty)$

Theorem 5. 21 Let S^∞ be a sequence of period 2^n. If $L(S^\infty) < 2^n$ and $0 \leqslant k \leqslant 2^{n-1}$, then

$$WC_{2k+1}(S^\infty) = 2^n$$

Proof: Let

$$w(x) = (\sum_{j=1}^{2k+1} x^{i_j})/(1 + x)^{2^s}$$
$$= v(x)/(1 + x)^{2^s}$$

Then w^∞ is periodic with period 2^n and $W_H(w^N) = 2k + 1$, where $N = 2^n$. Conversely, every binary sequence with period 2^n and Hamming weight $2k + 1$ can be expressed as the above $w(x)$, since $2k + 1$ is odd, so $gcd(v(x), (1+x)^{2^s}) = 1$. Therefore, we get $f_w(x) = (1+x)^{2^s}$ and $r_w(x) = v(x)$. It follows from Theorem 5. 6 that

$$WC_{2k+1}(S^\infty) = \min_{\substack{0 \leqslant i_j \leqslant 2^s-1 \\ i_j \neq i_s}} deg[(1 + x)^{2^s}/gcd[(1 + x)^{2^s},$$
$$r_w(x) + r_s(x)(1 + x)^{2^s-L(S^\infty)}]]$$

Noticing that $2^n - L(S^\infty) > 0$, we get from Lemma 5. 10 that

$$gcd(r_w(x) + r_s(x)(1 + x)^{2^s-L(S^\infty)}, (1 + x)^{2^s}) = 1$$

Thus, $WC_{2k+1}(S^\infty) = 2^n$. \qquad\qquad \#

Theorem 5. 22 Let S^∞ be a sequence of period 2^n. If $L(S^\infty) = 2^n$ and $k > 0$, then $WC_{2k}(S^\infty) = 2^n$.

Proof: Let w^∞ be a sequence of period 2^n and $W_H(w^N) = 2k$, where $N = 2^n$. Since $W_H(w^N)$ is even, $w^N(x)$ has root 1. Therefore $deg(f_w(x)) < 2^n$. Consequently, we have

$$gcd[(1 + x)^{2^s}, (1 + x)^{2^s-deg(f_w(x))}r_w(x) + r_s(x)] = 1$$

Hence, it follows from Theorem 5. 6 that $WC_{2k}(S^{\infty}) = 2^n$.

Theorem 5. 23 Let s^{∞} be a sequence of period 2^n. If $L(s^{\infty}) = 2^n$, $0 \leqslant k < 2^{n-1}$, let

$$V = \{t^N : L(t^{\infty}) = 2^n, W_H(s^N + t^N) = 2k\}$$

where $N = 2^n$. Then

$$WC_{2k+1}(s^{\infty}) \geqslant \min_{t^N \in V} WC_1(t^{\infty})$$

Proof: By definition, we have

$$WC_{2k+1}(s^{\infty}) = \min_{W_H(w^N) = 2k+1} L(s^{\infty} + t^{\infty}) \qquad (15)$$

Suppose w^{∞} is the sequence which makes the right side of (15) achieve its minimal value. Because $W_H(w^N) = 2k + 1 > 0$, assume $w_p = 1$, set $t^N(x) = x^p/(1+x)^{2^n}$ and $v^N(x) = w^N(x) + t^N(x)$, then $W_H(v^N) = 2k$ and $W_H(t^N) = 1$. By Theorem 5. 22 we know that $L(s^{\infty} + v^{\infty}) = 2^n$. Hence,

$$
\begin{aligned}
WC_{2k+1}(s^{\infty}) &= L(s^{\infty} + w^{\infty}) \\
&= L(s^{\infty} + v^{\infty} + t^{\infty}) \\
&\geqslant \min_{z \in V} WC_1(z^{\infty}) \qquad \#
\end{aligned}
$$

5. 3   Lower Bounds on the Weight Complexity of Binary ML-Sequences

A binary sequence is called an ML-sequence if it is generated by an n-stage LFSR that has period $2^n - 1$. Linear shift registers that generate ML-sequences are often employed as driving machines for running key generators, since ML-sequences have some good statistical properties and the linear complexity of the output sequences of these key stream generators is relatively easy to control. Because ML-sequences have a special position in stream ciphers, it is necessary to make clear the stability of linear complexity of ML-sequences. As binary sequences are practically used in stream ciphers, we only discuss the stability of linear complexity of binary ML-sequences in this section.

Lemma 5. 24   Let $g(x)$ be an irreducible polynomial with non-zero derivative $g'(x)$ over some field F , then $g(x)$ is a repeated factor of the polynomial $f(x)$ iff $g(x)$ divides $gcd(f(x), f'(x))$.

Proof: Let $g(x)$ be an irreducible factor of $f(x)$, so that $f(x) = g(x)h(x)$. Taking formal derivatives gives $f'(x) = g'(x)h(x) + g(x)h'(x)$. If $g(x)$ divides $f'(x)$ then it also divides $g'(x)h(x)$. But $deg(g'(x)) < deg(g(x))$, so $g(x)$ cannot divide $g'(x)$. In order to divide $f'(x)$, $g(x)$ must divide $h(x)$. Therefore $g(x)$ is a repeated factor of $f(x)$ iff it is a factor of both $f(x)$ and $f'(x)$, which happens iff $g(x)$ divides $gcd(f(x),\ f'(x))$.

Lemma 5. 25   In $GF(p)$ with p prime, every irreducible polynomial has nonzero derivative.

Proof: Let $f(x) = a_0 + a_1x + \cdots + a_nx^n$, an irreducible polynomial over $GF(p)$. Suppose that $f'(x) = 0$, then a_i must be equal to zero if p does not divide i , for $0 \leqslant i \leqslant n$. Therefore,

$$f(x) = a_0 + a_px^p + a_{2p}x^{2p} + \cdots$$

Hence, we have

$$f(x) = (a_0 + a_px + a_{2p}x^2 + \cdots)^p$$

This shows that $f(x)$ is reducible over $GF(p)$, a contradiction.

#

Theorem 5. 26   Let s^∞ be a binary ML-sequence with minimal polynomial f_s . If $deg(f_s) = m$, then

$$WC_1(s^\infty) \geqslant 2^m - m - 1$$

Proof: Let $N = 2^m - 1$, then the period of s^∞ is N . Set $g(x) = (x^N + 1)/f_s(x)$, since $f_s(x)$ is irreducible, so if $h(x)$ divides $x^N + 1$ and $h(x) \neq f_s(x)$, then $h(x)$ divides $y(x)$. Therefore, we get

$$gcd(x^N + 1,\ x^i + r_s(x)g(x)) = f_s^e(x), \qquad e > 0$$

On the other hand, it follows from Lemma 5. 25 that $f'_s(x) \neq 0$. Noticing $gcd(x^N + 1,\ x^{N-1}) = 1$, it follows from Lemma 5. 24 that $0 \leqslant e \leqslant 1$. Hence, by Lemma 5. 6 we obtain

$$WC_1(s^\infty) \geqslant 2^m - m - 1$$

#

It is well known that the number of binary ML-sequences of period $2^m - 1$ is $\Phi(2^m - 1)/m$, where $\Phi(\cdot)$ is the Euler function. Theorem 5. 26 tells us that one bit of change in every corresponding positions of every period will make the linear complexity of such a sequence jump to at least $2^m - m - 1$. This implies that the stability of linear complexity of nonlinear state filtered or nonlinear combined ML-sequence could be very bad. The conclusion has already been verified by the example given in Section 3. 3. On the other hand, a lot of binary sequences of period $2^m - 1$ can be obtained by changing one bit in each period in the corresponding places of a ML-sequence. From the viewpoint of linear complexity and the distribution of 0-runs and 1-runs, such a sequence is good. But from the viewpoint of stability of linear complexity, it is not a good one.

Lemma 5. 27   If the integer d is the greatest common divisor of the positive integer m and n , then the polynomial $x^d - 1$ is the greatest common divisor of the polynomial $x^m - 1$ and $x^n - 1$ over $GF(q)$.

Proof: Let $m = dm_1$, $n = dn_1$, then $gcd(m_1, n_1) = 1$, Noticing $x^m - 1 = (x^d - 1)(y^{m_1-1} + y^{m_1-2} + \cdots + y + 1)$ and $x^n - 1 = (x^d - 1)(y^{n_1-1} + y^{n_1-2} + \cdots + y + 1)$, where $y = x^d$, we see that $x^d - 1$ is a common divisor of $x^m - 1$ and $x^n - 1$ over $GF(q)$. To prove the lemma, it suffices to show that $y^{m_1} - 1 + \cdots + y + 1$ and $y^{n_1-1} + \cdots + y + 1$ have no common divisor over $GF(q)$. This is equivalent to showing that $y^{m_1} - 1$ and $y^{n_1} - 1$ has no common divisor other than $(y - 1)$ over $GF(q)$. Suppose that $y^{m_1} - 1$ and $y^{n_1} - 1$ have common divisor other than $y - 1$. Then $y^{m_1} - 1$ and $y^{n_1} - 1$ must have a common root a in some extension field of $GF(q)$, for instance, the splitting field of $(y^{m_1} - 1)(y^{n_1} - 1)$ with $a \neq 1$. Since $gcd(m_1, n_1) = 1$, there must exist two integers u and v such that $um_1 + vn_1 = 1$. Therefore, $a = (a^{m_1})^u \cdot (a^{n_1})^v = 1$, a contradiction. Hence $x^d - 1 = gcd(x^m - 1, x^n - 1)$.

#

Theorem 5. 28   Let s^∞ be a binary ML-sequence with minimal polynomial f_s of degree m . Denoting the maximum proper factor of $2^m - 1$ as

M , then

$$WC_2(s^\infty) \geqslant 2^m - m - M - 1.$$

Proof: Let $N = 2^m - 1$, w^∞ a binary sequence of period N with $W_H(w^N) = 2$. Suppose $w_i = w_j = 1$, $i < j$ and $w_k = 0$ if $k \neq i,j$. Setting $G = gcd(j - i, N)$, we get from Lemma 5. 27 that

$$f_w(x) = (x^N + 1)/(x^G + 1)$$

and

$$r_w(x) = (x^i + x^j)/(x^G + 1)$$

Since $f_s(x)$ is primitive and $G < N$, $f_s(x)$ does not divide $x^G + 1$. Therefore $f_s(x)$ divides $f_w(x)$. On the other hand, if $h(x)$ is an irreducible factor of $gcd(f_w(x), r_w(x) + r_s(x)f_w(x)/f_s(x))$, and $h(x) \neq f_s(x)$, then $h(x)$ divides $f_w(x)$ and $f_w(x)/f_s(x)$. Whence $h(x)$ divides $r_w(x)$, and $gcd(f_w(x), r_w(x))$. Noticing $gcd(f_w(x), r_w(x)) = 1$, we get $h(x) = 1$ and

$$gcd[f_w(x), r_w(x) + r_s(x)f_w(x)/f_s(x)] = f_s^e(x), \qquad e \geqslant 0$$

Because $x^N + 1$ has no repeated factor, neither does $f_w(x)$. Hence $0 \leqslant e \leqslant 1$. Whence it follows that

$$\begin{aligned} WC_2(s^\infty) &= \min_{0 \leqslant i < j \leqslant N-1} deg[f_s(x)f_w(x)/gcd(f_s(x)f_w(x), f_s(x)r_w(x) \\ &\quad + r_s(x)f_w(x))] \\ &= N - \max_{0 \leqslant i < j \leqslant N-1} deg[(x^G + 1)gcd(f_w(x), r_w(x) \\ &\quad + r_s(x)f_w(x)/f_s(x))] \\ &\geqslant N - (M + deg(f_s(x))) \\ &= 2^m - M - m - 1. \qquad\qquad\qquad \# \end{aligned}$$

Corollary 5. 28 Let s^∞ be a binary ML-sequence of period $2^m - 1$. If $2^m - 1$ is prime, then

$$WC_2(s^\infty) \geqslant 2^m - m - 2$$

Proof: Corollary 5. 28 is the special case $M = 1$, of Theorem 5. 27. The conclusion is apparently true.   #

Theorem 5. 29 Let s^∞ be a binary sequence of period $2^m - 1$, and $k < \lceil (2^m - 1)/m \rceil$. Then

$$WC_k(s^\infty) \geqslant \lceil (2^m - 1)/k \rceil - m$$

where $\lceil x \rceil$ here and hereafter denotes the smallest integer greater than or equal to x.

Proof: Let $N = 2^m - 1$ and let w^∞ be a binary sequence of period N with $W_H(w^N) = k$. Noticing that $f_s(x)$ is irreducible and $f_w(x)$ divides $x^N + 1$, we know that either $f_s(x)$ divides $f_w(x)$ or $gcd(f_s(x), f_w(x)) = 1$. Let us assume that $f_s(x)$ divides $f_w(x)$. Since $x^N + 1$ has no repeated factor, so does $f_w(x)$. It follows that

$$gcd(f_s(x)f_w(x), \; f_s(x)r_w(x) + r_s(x)f_w(x)) = f_s^e(x), \quad 0 \leqslant e \leqslant 2$$

whence

$$deg[f_s(x)f_w(x)/gcd(f_s(x)f_w(x), f_s(x)r_w(x) + r_s(x)f_w(x))]$$
$$\geqslant deg(f_w(x)) - deg(f_s(x)) \tag{16}$$

If $gcd(f_s(x), \; f_w(x)) = 1$, then $gcd(f_s(x)f_w(x), f_s(x)r_w(x) + r_s(x)f_w(x)) = 1$. Therefore, we have

$$deg[f_s(x)f_w(x)/gcd(f_s(x)f_w(x), \; f_s(x)r_w(x) + r_s(x)f_w(x))]$$
$$= deg(f_w(x)) + deg(f_s(x)) \tag{17}$$

Hence, it follows from Lemma 5. 6 and (16) as well as (17) that

$$WC_k(S^\infty)$$
$$\geqslant min\{ \min_{f_s(x)|f_w(x)} deg(f_w(x)) - m, \quad \min_{(f_s(x),f_w(x))=1} deg(f_w(x)) + m\}$$

On the other hand, since $W_H(w^N) = k$, there must exist an 0-run of length at least $\lceil (N - k)/k \rceil$ in the sequence w^∞. Thus, it follows from the basic property of linear complexity that

$$L(w^\infty) \geqslant \lceil (N - k)/k \rceil + 1 = \lceil N/k \rceil$$

This means that $deg(f_w(x)) \geqslant \lceil N/k \rceil$. Hence, we get

$$WC_k(s^\infty) \geqslant \lceil (2^m - 1)/k \rceil - m \qquad \#$$

We would now like to mention that the bounds given in Theorem 5. 27 and Corollary 5. 28 are very tight. For small k, the bound given in Theorem 5. 29 is also tight, but may not be tight for large k. Another thing we want to make clear is that one must be careful when employing LFSRs as driving machines for stream ciphers, since Theorems 5. 27 and 5. 29 tell us again that to make the linear complexity of a sequence based on ML-sequences large enough is easy, but the stability of linear complexity of the produced sequence is really bad.

5. 4 Lower Bounds on the Linear Complexity of Nonlinear Filtered ML—Sequences Derived from That of Weight Complexity

Nonlinear state filtered LFSRs are usually used as key stream generators, as depicted in Fig. 3. 1. Compared with the nonlinear combiner driven by LFSRs, the linear complexity of the output sequences of nonlinear state filtered LFSRs is rather more difficult to control. For the former, Rueppel and Staffelback [Ruep88] proved that if the lengths of the driving maximum-length LFSRs are prime to each other, assume that the combining function is

$$f(x_1, x_2, \cdots, x_n) = a_0 + \sum x_i a_i + \sum a_{ij} x_i x_j + \cdots$$
$$+ a_{1 \cdot 2 \cdots n} x_1 x_2 \cdots x_n, \ a_i, a_{ij}, \cdots, \in GF(q)$$

Then the linear complexity of the produced sequence is $F(L_1, L_2, \cdots, L_n)$, where $L_1, L_2, \cdots, L_n$ denote the lengths of the driving maximum-length LFSRs, and $F(x_1, x_2, \cdots, x_n)$ is defined as

$$F(x_1, x_2, \cdots, x_n) = a_0^1 + \sum a_i^1 x_i + \sum a_{ij}^1 x_i x_j$$
$$+ \cdots + a_{1 \cdot 2 \cdots n}^1 x_1 x_2 \cdots x_n$$

with a_i^1, $a_{ij}^1 \cdots$, being 1 if a_i, a_{ij}, $\cdots$, are nonzero, and zero otherwise, and $F(x)$ is evaluated over integers and not in $GF(q)$. This demonstrates that the linear complexity of the output sequences of the nonlinear combiner is controllable. For the linear complexity of the output sequences of the nonlinear state filtered LFSRs, Key [Key 76] developed a general upper bound, i. e. , if the nonlinear order of the filter function is k, then the linear complexity of the output sequence is upper bounded by $\sum_{i=1}^{k} \binom{n}{i}$, where n denotes the length of the driving LFSR. But in cryptographic applications one is rather interested in generating sequences with a guaranteed large minmum linear complexity, Kumar and Scholtz [Kuma 83] derived a lower bound which slightly exceeds

$(^{L}_{L/2})2^{L/4}$ for the class of bent sequences, where L , denoting the length of an ML-sequence, is restricted to be a multiple of 4. Rueppel [Ruep 86] and Bernasconi and Günter [Bern 85] derived a general lower bound of $(^{L}_{k})$ on the linear complexity of nonlinear filtered ML-sequences which holds for a broad and simple class of functions, and is not constrained in the length L of the maximum-length LFSR or the nonlinear order k of the function employed. We now derived a general bound on the nonlinear state filtered ML-sequences based on that of weight complexity of ML-sequences, which holds for every nonlinear state filtered ML-sequence.

Assume a Boolean function $f : GF(2)^m \rightarrow GF(2)$ is applied to a maximum length LFSR of length n with $n \geqslant m$, and f taps at the i_1th , i_2th , $\cdots$, i_mth storage cells of the LFSR. We can regard $f(x)$ as taping at all the storage cells by introducing a new filter function

$$g(x_1, x_2, \cdots, x_n) = f(x_{i_1}, x_{i_2}, \cdots, x_{i_m})$$

where the new input variable x_i takes its values from the ith storage cell. Let s^∞ and z^∞ be the output sequences of the driving maximum-length LFSR and the nonlinear state filtered maximum—length LFSR respectively. Let $a = \max\limits_{w} |S_{(g)}(w)|$ and the best affine approximation of g (x) be $L + wx$, where $L = 0$ if $S_{(g)}(w) \geqslant 0$; otherwise $L = 1$. Let t^∞ be the output sequence of the wx filtered LFSR, where the LFSR is the same as the original maximum-length LFSR. Then t^∞ is also a ML-sequence (m-sequence or PN-sequence). If $L = 1$, the number of agreements between $L \oplus wx$ and $g(x)$ is $2^{n-1} + 2^{n-1}a - g(0)$ as x runs over $GF(2)^n - \{0\}$. Therefore the number of disagreements between wx and $\bar{g}(x)(= g(x) + 1)$ as x runs over $GF(2)^n - \{0\}$, is $2^n - 1 - (2^{n-1} + 2^{n-1}a - g(0)) = 2^{n-1} - 2^{n-1}a + g(0) - 1$. Thus the number of disagreements between t^N and $\bar{z}^N$ is $2^{n-1} - 2^{n-1}a + g(0) - 1$, where $N = 2^n - 1$. Noticing that t^∞ is a maximum-length sequence, it follows from Theorem 5. 29 that

$$L(\bar{z}^\infty) \geqslant \lceil (2^n - 1)/(2^{n-1} - 2^{n-1}a + g(0) - 1) \rceil - n$$

Hence, it follows from Lemma 3. 5 that

$$L(z^\infty) \geqslant \lceil (2^n - 1)/(2^{n-1} - 2^{n-1}a + g(0) + 1) \rceil - n - 1$$

$$(18)$$

If $L = 0$, the number of agreements between wx and $g(x)$ is $2^{n-1} + 2^{n-1}a + g(0) - 1$. Consequently, the number of disagreements between s^N and z^N is $2^{n-1} - 2^{n-1}a - g(0) + 1$. It follows from Theorem 5. 29 that

$$L(z^\infty) \geqslant \lceil (2^n - 1)/(2^{n-1} - 2^{n-1}a - g(0) + 1) \rceil - n \quad (19)$$

Combining (18) and (19). we obtain the following result:

Theorem 5. 30 Let z^∞ be a nonlinear state filtered binary ML-sequence of period $2^n - 1$, $f(x_1, x_2, \cdots, x_m)$ be the nonlinear filter function with tap positions $i_1, i_2, \cdots, i_m$ respectively. Then

$$L(z^\infty) \geqslant \lceil (2^n - 1)/(2^{n-1} - 2^{n-1}a + |g(0) - 1|) \rceil - n - 1$$

$$(20)$$

where

$$g(x_1, x_2, \cdots, x_n) = f(x_{i_1}, x_{i_2}, \cdots, x_{i_m})$$

$$a = \max_w |S_{(g)}(w)| \qquad\qquad \#$$

Example 1: Let the nonlinear function $x_1 + x_2 + \cdots + x_{n-1} + x_1 x_2 \cdots x_{n-1}$ filter a maximum-length LFSR of length n with tap position $1, 2, \cdots, n$ -1. then $g(x_1, x_2, \cdots, x_n) = f(x_1, x_2, \cdots, x_{n-1}) = x_1 + x_2 + \cdots + x_{n-1}$ $+ x_1 \cdot x_2 \cdots x_{n-1}$. By calculation, we have

$$a = \frac{2^n - 4}{2^n}$$

and

$$L(z^\infty) \geqslant \lceil \frac{2^n - 1}{3} \rceil - n - 1$$

Remark : From (12) we get that

$$L(z^\infty) \geqslant \lceil (2^n - 1)/2^{n-1}(1 - a) \rceil \quad n \quad 1.$$

Thus the bound given in (20) is tight for filter functions with large a , bad for those with small a .

5.5 Lower Bounds on the Weight Complexity of Clock-Controlled Binary Sequences

Clock controlled sequences employed as running key sequences were investigated by T. Beth and F. C. Piper [Beth. 84], Kjeldsen Andresen [Andr 80], Rainer Vogel [Voge 84], Dieter Gollmann [Goll 84], C. G. Günther [Günt 87] and Rainer A. Rueppel [Ruep 87], etc. The general results about clock controlled shift- register sequences are not easy to obtain, but they are clear for some special cases. The linear complexity of output sequence of the following stop-and-go generator is known.

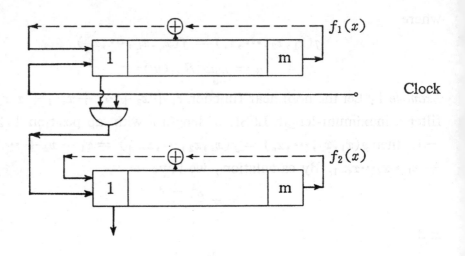

Fig 5.2 The stop-and-go generator [Beth 84]

Whence the output sequence of SRG1 is (a_i), and that of SRG2 is (b_i), and

$$G(0) = 0, \qquad G(k) = \sum_{i < k} a_i$$

We now present the result of linear complexity of the output sequence of the stop-and-go generator obtained by T. Beth [Beth 84], R. Vogel [Voge 84].

Lemma 5. 31   Let $f_1(x)$, $f_2(x)$, $\cdots$, $f_N(x)$ be the set of all irreducible polynomials of degree m and order (or exponent) e with

$$e = (2^m - 1)/d$$

and let λ be a positive integer with the properties

i) $(\lambda, d) = 1$

ii) All prime divisors of λ are prime divisors of $2^m - 1$, then

a) $\lambda \cdot m$ is the least number with the property $\lambda \cdot e \mid 2^{\lambda \cdot m} - 1$

b) The polynomials $f_1(x^\lambda)$, $\cdots$, $f_N(x^\lambda)$ are irreducible of degree $m \cdot \lambda$ and order $\lambda \cdot e$ and the set $\{f_1(x^\lambda), \cdots, f_N(x^\lambda)\}$ consists of all polynomials with these properties.

Proof [Dick oo]:

In the special case $e = \lambda = 2^m - 1$ and $d = 1$, we get

a) $m(2^m - 1) = min\{k : (2^m - 1)^2 \mid 2^k - 1\}$

b) The set $\{f_1(x^{2^m-1}), f_2(x^{2^m-1}), \cdots, f_N(x^{2^m-1})\}$

is the set of all polynomials of degree $m(2^m - 1)$ with order $(2^m - 1)^2$.

With the help of this theorem we now prove the following lemma.

Lemma 5. 32   Let the feedback polynomials $f_1(x)$ and $f_2(x)$ be primitive of degree m . Then

1) $f_2(x^{2^m-1})$ is the minimal polynomial of the sequence $(b_{G(k)})$. Thus the linear complexity of $(b_{G(k)})$ is $m(2^m - 1)$.

2) The sequence $(b_{G(k)})$ has the minimum period $(2^m - 1)^2$.

Proof : [Voge 84] As already shown, since the order of $f_2(x^{2^m-1})$ is $(2^m - 1)^2$, the minimum period of $(b_{G(k)})$ is $(2^m - 1)^2$.

1) Setting $p = 2^m - 1$, the sum has the properties

$$G(k + n \cdot p) = G(k) + n \cdot G(p)$$

$$G(p) = 2^{m-1}$$

Denoting E as the shift operator and using the operator $f(E^p)$ corresponding to the polynomial $f(x^p)$, we get

$$f_2(E^p)b_{G(k)} = \sum_{i=0}^{m} c_i \cdot b_{G(k+ip)}$$

$$= \sum_{i=0}^{m} c_i \cdot b_{G(k)+iG(p)}$$

$$= \sum_{i=0}^{m} c_i b_{G(k)+i2^{m-1}}$$

$$= \sum_{i=0}^{m} c_i b_{k'+i2^{m-1}}$$

$$= f_2(E^{2^{m-1}})b_{k'}$$

$$= (f_2(E))^{2^{m-1}} b_{k'}$$

$$= 0$$

where $k' = G(k)$, and c_i are the coefficients of $f_2(x)$. Since $f_2(x^{2^m-1})$ is irreducible, it is the minimal polynomial of $(b_{G(k)})$.

With the help of Lemma 5. 32, we now develop lower bounds on the weight complexity of the output sequence of the stop-and-go generator.

Theorem 5. 33 Let the feedback polynomials $f_1(x)$ and $f_2(x)$ in Fig. 5. 2 be primitive of degree m . Denote the output sequence $(b_{G(k)})$ as s^∞ . Then

$$1) WC_1(s^\infty) \geqslant (2^m - 1)(2^m - m - 1)$$
$$2) WC_2(s^\infty) \geqslant (2^m - 1)(2^m - M - m - 1)$$

Proof: 1) Let $N = (2^m - 1)^2$, then it follows from Lemma 5. 24 that $x^N + 1$ has no repeated factor. Lemma 5. 32 tells us that $f_s(x) = f_2(x^{2^m-1})$. If $h(x)$ divides $x^N + 1$, and $f_s(x)$ does not divide $h(x)$, then $h(x)$ must divide $(x^N + 1)/f_s(x)$. Therefore,

$$gcd(x^N + 1, \, x^i + r_s(x)g(x)) = f_2(x^{2^m-1})^e, \, 0 \leqslant e \leqslant 1$$

where $r_s(x)$ and $g(x)$ are defined as in Theorem 5. 6. It follows from Theorem 5. 6 that

$$WC_1(s^\infty) \geqslant (2^m - 1)^2 - m(2^m - 1)$$
$$= (2^m - 1)(2^m - m - 1)$$

2) Let w^∞ be a binary sequence of period N , and $W_H(w^N) = 2$

with $w_i = w_j = 1$, $w_k = 0$ for $k \neq i, j$. Let $G = gcd\ (|j - i|, N)$ then by Theorem 5. 3 we get

$$f_w(x) = (x^N + 1)/(x^G + 1)$$
$$r_w(x) = (x^i + x^j)/(x_G + 1)$$

Noticing the order of $f_2(x^{2^m-1})$ is N and $G < N$, We know that $f_2(x^{2^m-1})$ does not divide $x^G + 1$. Consequently, it must divide $f_w(x)$. On the other hand, $f_w(x)$ has no repeated factor (since $x^N + 1$ has no repeated factor) and $gcd\ (f_w(x), r_w(x)) = 1$. It follows that

$$gcd(f_w(x), r_w(x) + r_s(x)f_w(x)/f_s(x))$$
$$= f_2(x^{2^m-1})^e \qquad\qquad 0 \leqslant e \leqslant 1$$

Hence, by Theorem 5. 6,

$$WC_2(s^\infty) = N - \max_{0 \leqslant i < j \leqslant N-1} deg[(x^G + 1)gcd(f_w(x), r_w(x)$$
$$+ r_s(x)f_w(x)/f_s(x))]$$
$$\geqslant N - [M(2^m - 1) + deg(f_2(x^{2^m-1}))]$$
$$= (2^m - 1)(2^m - M - m - 1) \qquad\qquad \#$$

Corollary 5. 34   If $2^m - 1$ is prime, then

$$WC_2(s^\infty) \geqslant (2^m - 1)(2^m - m - 2) \qquad\qquad \#$$

Theorem 5. 35   Let s^∞ be a sequence of period N. If the minimal polynomial of s^∞ is $f_s(x)$, then

$$WC_k(s^\infty) \geqslant \lceil N/k \rceil - deg(f_s(x))$$

and the bound is also suitable for $SC_k(s^\infty)$.

Proof: We first prove that $gcd(g(x)h(x), f(x))$ divides $gcd\ (g(x), f(x))gcd(h(x), f(x))$ for any polynomials $g(x)$, $h(x)$ and $f(x)$ over any field. Noticing that

$$gcd(g(x)h(x), f(x)) = gcd(g(x), f(x)) \cdot gcd$$
$$[h(x), f(x)/gcd(g(x), f(x))]$$

and $gcd[h(x), f(x)/gcd(g(x), f(x))]$ divides $gcd(h(x), f(x))$, we get that $gcd(g(x)h(x), f(x))$ divides $gcd(g(x), f(x))gcd(h(x), f(x))$. Let $r_s(x)/f_s(x)$ and $r_w(x)/f_w(x)$ be the reduced rational forms of s^∞ and w^∞ respectively, where w^∞ is a sequence with the same period as s^∞ and $W_H(w^N) = k$, then by Theorem 5. 6 we get

$$WC_k(s^\infty) = \min_{W_H(w^N)=k} deg[f_s(x)f_w(x)/gcd(f_s(x)f_w(x),\ r_s(x)f_w(x) +$$

$$f_s(x)r_w(x))]$$

$$= \min_{W_H(w^N)=k} [deg(f_s(x)) + deg(f_w(x))$$

$$- deg(gcd(f_s(x)f_w(x),\ f_s(x)r_w(x) + r_s(x)f_w(x)))]$$

$$\geqslant \min_{W_H(w^N)=k} [deg(f_s(x)) + deg(f_w(x))$$

$$- deg(gcd(f_s(x),f_s(x)r_w(x) + r_s(x)f_w(x)))$$

$$- deg(gcd(f_w(x),f_s(x)r_w(x) + r_s(x)f_w(x)))]$$

$$= \min_{W_H(w^N)=k} [deg(f_s(x)) + deg(f_w(x))$$

$$- 2deg(gcd(f_s(x),f_w(x)))]$$

$$\geqslant \min_{W_H(w^N)=k} [deg(f_w(x)) + deg(f_w(x)) - 2deg(f_s(x))]$$

$$\geqslant \min_{W_H(w^N)=k} [deg(f_w(x)) - deg(f_s(x))]$$

$$= \min_{W_H(w^N)=k} deg(f_w(x)) - deg(f_s(x))$$

Since the period of w^∞ is N and $W_H(w^N) = k$, there must exist an 0-run of length at least $\lceil(N-k)/k\rceil$. Therefore the linear complexity of w^∞ is greater than or equal to $\lceil(N-k)/k\rceil + 1 = \lceil N/k\rceil$. It follows that $deg(f_w(x)) \geqslant \lceil N/k\rceil$, and

$$WC_k(s^\infty) \geqslant \lceil N/k\rceil - deg(f_s(x)) \qquad \#$$

Corollary 5. 36 Let the assumptions about the generator in Fig. 5. 2 be as in Theorem 5. 33, then

$$WC_k(s^\infty) \geqslant \lceil(2^m - 1)^2/k\rceil - m(2^m - 1) \qquad \#$$

Two types of pseudorandom number generators based on linear feedback shift registers have been investigated intensively in the literature. The first type of generator is based on the nonlinear feedforward combination of the output of one or several LFSRs [Sieg 84][Ruep 88] and the second one is based on a chain of LFSRs in which the clock of a given LFSR is controlled by the output of the preceding LFSR in that chain [Andr 80, Voge 84, Goll 84, Beth 84, Cham 84]. The generators of the first type can be chosen to have a very large period and excellent statistical properties and to be secure against a correlation attack on one or several LFSRs. These generators, however, have a linear complexity that only depends algebraically on the length of the various LFSRs. The generators of the second type can be chosen to have a very large period and a linear complexity that is an exponential function of the length of one LFSR. Depending on the exact specifications of the generators, one will, however, either have bad statistics (stop-and-go generator) [Voge 84] or a reduced bits rate (binary rate multiplier) [Cham 84]. Furthermore, in both cases the generator shows some cryptographic weaknesses. Based on the strengths and weaknesses of both schemes, Günther proposed the type of generator [Günt 87] shown in Fig. 5. 3, where SR1, SR2 and $\overline{SR2}$ are LFSRs, and the feedback polynomial of $\overline{SR2}$ is the reciprocal one of SR2s.

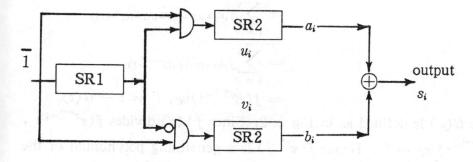

Fig 5. 3 Günther's key stream generator

In order to investigate the stability of linear complexity of the output sequences of the generators, we need the following lemma obtained by Wu [Wu 87].

Lemma 5. 37 Let the feedback polynomials of SR1 and SR2 be primitive and be respectively $g(x)$ and $f(x)$ with degree m, then the output sequence of the key stream generator in Fig 5. 3 has period $(2^m - 1)^2$ and linear complexity $m(2^m - 1)$.

Proof [Wu 87]: By assumption, we see that the feedback polynomial of $\overline{SR2}$ is $f^*(x)$, the reciprocal polynomial of $f(x)$. We start by proving that $f^*(x)$ divides $f(x^{2^{m-1}-1})$. If α is a root of $f^*(x)$, then α is primitive of order $2^m - 1$, and α^{-1} is a root of $f(x)$. So $f(\alpha^{-1}) = 0$. Noticing $\alpha^{2^m-1} = 1$, we get $\alpha^{2^{m-1}-1} = (\alpha^{-1})^{2^{m-1}}$. Therefore

$$f(\alpha^{2^{m-1}-1}) = (f(\alpha^{-1}))^{2^{m-1}} = 0$$

Thus every root of $f^*(x)$ is one of $f(x^{2^{m-1}-1})$. On the other hand, since $x^{2^m-1} - 1$ has no repeated factor, neither does $f^*(x)$. Hence, $f^*(x)$ divides $f(x^{2^{m-1}-1})$. Let u^∞ and v^∞ be the output sequences of $SR2$ and $\overline{SR2}$ respectively, then $a_t = u_{G(t)}$, $b_t = v_{t-G(t)}$. Let

$$f(x) = \sum_{i=0}^{m} f_i x^i$$

and E be the shift operator, then we have

$$f(E^{2^m-1})b_t = \sum_{i=0}^{m} f_i b_{t+i(2^m-1)}$$

$$= \sum_{i=0}^{m} f_i u_{t+i(2^m-1)-G(t+i(2^m-1))}$$

$$= \sum_{i=0}^{m} f_i v_{t-G(t)+i(2^{m-1}-1)}$$

$$= f(E^{2^{m-1}-1})v_{t'}, \quad t' = t - G(t)$$

where $G(.)$ is defined as in Fig 5.2. Since $f^*(x)$ divides $f(x^{2^{m-1}-1})$, $f(E^{2^{m-1}-1})v_{t'} = 0$. Hence $f(x^{2^m-1})$ is a generating polynomial of the sequence b^∞. It follows from Lemma 5.31 that $f(x^{2^m-1})$ is irreducible, so $f(x^{2^m-1})$ is the minimal polynomial of b^∞. Lemma 5.31 also tell us that the order of $f(x^{2^m-1})$ is $(2^m - 1)^2$, so the period of b^∞ is $(2^m - 1)^2$. Lemma 5.32 shows that the sequence a^∞ has minimal polynomial $f(x^{2^m-1})$ and period $(2^m - 1)^2$. Let the reduced rational form of $a^\infty(x)$ and $b^\infty(x)$ be respectively

$$a^\infty(x) = r_a(x)/f(x^{2^m-1})$$
$$b^\infty(x) = r_b(x)/f(x^{2^m-1})$$

Noticing that

$$(a+b)^\infty(x) = (r_a(x) + r_b(x)/f(x^{2^m-1}))$$

and that $f(x^{2^m-1})$ is irreducible, we see that the minimal polynomial of s^∞ is $f(x^{2^m-1})$ provided that $a^m \neq b^m$. Thus, s^∞ has period $(2^m - 1)^2$ and minimal polynomial $f(x^{2^m-1})$. Because the degree of $f(x^{2^m-1})$ is $m(2^m - 1)$, s^∞ has linear complexity $m(2^m - 1)$.   #

It follows from Lemma 5.37, Theorems 5.33 and 5.35, and Corollary 5.36 that the lower bounds on the complexity of the output sequences of the stop-and-go generator are also satisfied by the output sequences of Günther's generator. It follows that $WC_k(s^\infty) \geqslant m(2^m - 1) = L(s^\infty)$ provided that $k \leqslant \lceil (2^m - 1)/2m \rceil$, where $\lceil x \rceil$ is the floor function, i. e. , the largest integer smaller than or equal to x. This shows that the linear-complexity stability of the foregoing two kinds of sequence is good to some extent from the cryptographic viewpoint. Due to the merits of Günther's generators we now consider a class of slightly different generators as shown in Fig. 5.4.

Before investigating the linear-complexity stability of the output sequence s^∞ produced by the generator in Fig 5.4, we need to analyze its minimal polynomial and period.

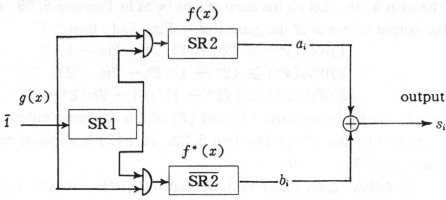

Fig 5.4 The modulo-2 addition of two related stop-and-go generators

Theorem 5.38   Let the feedback polynomials $f(x)$, $g(x)$ be primitive over $GF(2)$ and of degree m. If the output sequence $s^\infty \neq 0$ and $f(x) \neq f^*(x)$, then it has linear complexity $2m(2^m - 1)$ and mininal polynomial $f^*(x^{2^m-1})f(x^{2^m-1})$ as well as period $(2^m - 1)^2$.

Proof: Noticing that the sequences a^∞ and b^∞ in Fig. 5. 4 are outputs of two stop-and-go generators, we get from Lemma 5. 32 that the reduced rational forms of $a^\infty(x)$ and $b^\infty(x)$ are respectively

$$a^\infty(x) = r_a(x)/f(x^{2^m-1})$$
$$b^\infty(x) = r_b(x)/f^*(x^{2^m-1})$$

Therefore

$$(a+b)^\infty(x) = \frac{r_a(x)f^*(x^{2^m-1}) + r_b(x)f(x^{2^m-1})}{f(x^{2^m-1})f^*(x^{2^m-1})}$$

Noticing that $gcd(r_a(x), f(x^{2^m-1})) = 1$, and $gcd(r_b(x), f^*(x^{2^m-1})) = 1$, we see that

$$gcd\left[r_a(x)f^*(x^{2^m-1}) + r_b(x)f(x^{2^m-1}),\right.$$
$$\left. f(x^{2^m-1})f^*(x^{2^m-1})\right] = 1$$

provided that $f(x) \neq f^*(x)$ and $s^\infty \neq 0$. Therefore, the minimal polynomial of s^∞ is $f(x^{2^m-1})f^*(s^{2^m-1})$ and s^∞ has linear complexity $2m(2^m - 1)$. Since $ord(f(x^{2^m-1}) = ord(f^*(x^{2^m-1})) = (2^m - 1)^2$ and $f(x) \neq f^*(x)$, $ord(f(x^{2^m-1})f^*(x^{2^m-1})) = (2^m - 1)^2$ #

Theorem 5. 39 Let all the assumptions be as in Theorem 5. 39, and s^∞ the output sequence of the generator in Fig. 5. 4, then

$$1) WC_1(s^\infty) \geqslant (2^m - 1)(2^m - 2m - 1)$$
$$2) WC_2(s^\infty) \geqslant (2^m - 1)(2^m - 2m - 2)$$
$$3) WC_k(s^\infty) \geqslant \lceil (2^m - 1)^2/k \rceil - 2m(2^m - 1)$$

Proof: We can prove parts (1) and (2) of this theorem similarly to the proof of (1) and (2) of Theorem 5. 33. Part (3) is a special result of Theorem 5. 35. #

It follows from Theorem 5. 39 that $WC_k(s^\infty) \geqslant L(s^\infty) = 2m(2^m - 1)$ for any positive integer k with $k < \lceil (2^m - 1)/4m \rceil$. Thus, the linear-complexity stability of the output sequence of the key stream generator is good to some degree from the practical viewpoint. It is evident that the values of linear complexity divided by period is the same for both the ML-sequences and the stop-and-go sequences. Noticing that the values of bounds of weight complexity divided by period for both ML-sequences and that of the clock-controlled sequences in this section, we conclude that the stability of ML-sequences and of clock-controlled sequences is theoretically and proportionally identical.

5. 6 A Lower Bound on the Linear Complexity of the Clock-Controlled and Nonlinear-Filtered Binary ML-Sequences

As already seen, clock-controlled ML-sequences have both large linear complexity and relatively good linear-complexity stability. But some of their statistical properties might be bad. Due to their merits and demerits, we can expect that the nonlinear filtered distinct phases of a clock-controlled ML-sequence could remain good properties, but eliminate or improve bad ones. Anyway, it is necessary to clarify the linear complexity of the output sequence of the key streamgenerator shown in Fig. 5. 5. Based on the results about the weight complexity of the clock controlled ML — sequences, we shall develop a lower bound on the linear complexity of the output sequences of the generator as follows. To analyze the key stream generator in Fig. 5. 5, we must understand how the state vector of LFSR2 changes. To do so, we need the following result about binary sequences.

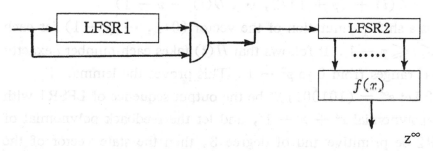

Fig. 5. 5 The clock controlled and nonlinear-filtered LFSR

Lemma 5. 40 Let s^∞ be a binary sequence with an odd period P , and

$$G(t) = \sum_{i<t} s_i \text{ with } G(0) = 0 \text{ and } G(p) = (p+1)/2 . \text{ Setting}$$

$$H(t) = G(t)\,(mod\,p)$$

then in the vector $(H(0), H(1), \cdots, H(p^2-1))$, there are exactly p components with value i for each $0 \leqslant i \leqslant p-1$.

Proof: Noticing that $G(k+np) = G(k) + nG(p)$ and $G(p) = (p+1)/2$, we can show that

$$(H(0),\ H(2p), \cdots,\ H((p-1)p),$$
$$H(p),\ H(3P), \cdots,\ H((p-2)p))$$
$$= (0, 1, \cdots,\ (p-1)/2,\ (p+1)/2, \cdots,\ p-1).$$

For every $0 \leqslant i \leqslant p-1$, by the assumptions we get

$$H(i+2kp) = (G(i) + 2kG(p))\ (mod\,p)$$
$$= (G(i) + 2k(p+1)/2)\ (mod\,p)$$
$$= (G(i) + k)\ (mod\,p)$$

for $0 \leqslant k \leqslant (p-1)/2$, and

$$H(i+(2k+1)/p) = (G(i) + (2k+1)G(p))\ (mod\,p)$$
$$= (G(i) + (2k+1)(p+1)/2)\ (mod\,p)$$
$$= (G(i) + k + (p+1)/2)\ (mod\,p)$$

for $0 \leqslant k \leqslant (p-3)/2$. Consequently.

$$y_i = (H(i), H(i+2p), \cdots,\ H(i+(p-1)p),$$
$$H(i+p), H(i+3p), \cdots,\ H(i+(p-2)p))$$
$$= (G(i), G(i)+1, \cdots, G(i)+(p-1)/2,$$
$$G(i)+(p+1)/2,\ \cdots,\ G(i)+p-1)$$

Thus, y_i is a shift permutation of the vector $(0, 1, \cdots, p-1)$ for each i with $0 \leqslant i \leqslant p-1$. It follows that $H(t)$ takes each number i exactly p times as t ranges from 0 to $p^2 - 1$. This proves the lemma. #

Example 2 Let $s^\infty = (1010011)^\infty$ be the output sequence of LFSR1 with feedback polynomial $x^3 + x + 1$, and let the feedback polynomial of the LFSR2 be primitive and of degree 3, then the state vector of the LFSR2 changes cyclicly in the following way:

$$S_0 S_1 S_1 S_2 S_2 S_2 S_3 S_4 S_5 S_5 S_6 S_6 S_6 S_0 S_1 S_2 S_2 S_3 S_3 S_3 S_4$$
$$S_5 S_6 S_6 S_0 S_0 S_0 S_1 S_2 S_3 S_3 S_4 S_4 S_4 S_5 S_6 S_0 S_0 S_1 S_1 S_1 S_2 S_3 S_4$$

$$S_4 S_5 S_5 S_5 S_6 \cdots$$

In the above vector sequence every S_i occurs exactly 7 times.

Theorem 5. 41 Let the feedback polynomial $g(x)$ and $f(x)$ of LFSR1 and LFSR2 be primitive and of degree n respectively, and let $f(x_1, x_2, \cdots, x_m)$ be the nonlinear filter function with tap position $i_1, i_2, \cdots, i_m$, $m \leqslant n$. Then

$$L(z^\infty) \geqslant min\{WC_k(t^\infty), WC_{k+2^n-1}(t^\infty), WC_{k-2^n-1}(t^\infty)\} - 1,$$

where $k = (2^n - 1)(2^{n-1} - 2^{n-1}a)$, and $a = max|S_{(g)}(w)|$ with $g(x_1, x_2, \cdots, x_n) = f(x_{i_1}, x_{i_2}, \cdots, x_{i_m})$, and t^∞ is the output sequence of the stop-and-go generator of Fig. 5. 5.

Proof: Let s^∞ and z^∞ be the output sequences of the stop-and-go generator and the generator in Fig. 5. 5. By the assumptions in the theorem, we see that the BAA of $g(x)$ is $L + wx$, where $L = 0$ if $S_{(g)}(w) \geqslant 0$; otherwise $L = 1$. Let t^∞ be the output sequence of the generator when the filter function is replaced by wx, then t^∞ is the modulo-2 addition of several distinct phases of s^∞, so it is also an output sequence of the stop-and-go generator in Fig. 5. 5 with only a different initial state of LFSR 2, and its minimal polynomial is $f(x^{2^2} - 1)$. If $L = 1$, the number of agreements between $L + wx$ and $\bar{g}(x)$ is $2^{n-1} + 2^{n-1}a - g(0)$ as x runs over $GF(2)^n - \{0\}$. Whence, the number of disagreements between $L + wx$ and $\bar{g}(x)$ is $2^{n-1} - 2^{n-1}a + g(0) - 1$. Thus, it follows from Lemma 5. 40 that the number of disagreements between t^N and $\bar{z}^N$ is $(2^n - 1)(2^{n-1} - 2^{n-1}a + g(0) - 1) = k + (2^n - 1)(g(0) - 1)$. By the definition of weight complexity, we have

$$L(\bar{z}^\infty) \geqslant WC_{k+(2^n-1)(g(0)-1)}(t^\infty)$$

Hence, it follows from Lemma 3. 5 that

$$L(z^\infty) \geqslant WC_{k+(2^n-1)(g(0)-1)}(t^\infty) - 1 \qquad (21)$$

If $L = 0$, similar to the case $L = 1$, we see that the number of agreements between t^N and z^N is $(2^n - 1)(2^{n-1} - 2^{n-1}a + 1 - g(0)) = k + (2^n - 1)(1 - g(0))$, where $N = (2^n - 1)^2$. By the definition of weight complexity, we have

$$L(z^\infty) \geqslant WC_{k+(2^n-1)(1-g(0))}(t^\infty) \qquad (22)$$

Combining (21) and (22), we obtain Theorem 5. 41.   #

One can get a good lower bound on the linear complexity of the output sequence of the generator in Fig. 5. 5. if he or she is able to develop a tight lower bound on the weight complexity of the output sequence of the stop-and-go generators. Unfortunately, the lower bound developed in Corollary 5. 36 is not ideal at all for large k. Consequently, we look forward to any tight bound on the weight complexity of the output sequence of the stop-and-go generator.

5. 7 Another Approach to the Stability of Linear Complexity of Sequences

Weight complexity and sphere complexity were introduced to measure the linear-complexity stability of sequences. Weight complexity $WC_k(s^\infty)$ measures how small the linear complexity can be made after changing k bits of s^∞ arbitrarily in every period in the corresponding places. The geometrical meaning of $WC_k(s^\infty)$ is the minimum linear complexity of sequences on a sphere surface in some Hamming space. Lower and upper bounds on the weight complexity of various kinds of sequences were developed in the foregoing sections for the purpose of estimating linear-complexity stability. To measure the stability of linear-complexity of sequences, we now consider another problem, i. e. , the determination of the sequence with linear complexity k and period N which is the "closest" one to a given sequence. To do so, let us first define two kinds of measure indexes.

Definition 5. 42   Let s^∞ be a sequence of period N over $GF(q)$. Its fixed-complexity distance (FCD) and variable-complexity distance (VCD) are respectively defined as

$$FCD_k(s^\infty) = \min_{\substack{L(t^\infty)=k \\ Per(t^\infty)=N}} W_H(s^N + t^N)$$

$$VCD_k(s^\infty) = \min_{\substack{0 < L(t^\infty) \leqslant k \\ Per(t^\infty) = N}} W_H(s^N + t^N)$$

where $Per(t^\infty)$ denotes the period of t^∞.

It follows from the above definition that $FCD_k(s^\infty)$ and $VCD_k(s^\infty)$ measure the extent to which s^∞ can be approximated by the set of sequences with linear complexity k and with linear complexity less than or equal to k respectively. In what follows, we shall investigate the relationships between weight complexity and fixed-complexity distance, as well as sphere complexity and variable-complexity distance.

5.7.1 The Relationships Between Weight Complexity and Fixed-Complexity Distance as well as Sphere Complexity and Variable-Complexity Distance

The connection between the Hamming weight of a finite sequence and the linear complexity of a related sequence was established by Blahut [Blah 79], Massey and Shaub applied it to coding theory and got excellent results about the minimum distance of cyclic codes [Mass 87]. Before performing the task of developing the relationships, we need the discrete Fourier transform (DFT) in a Galois field and Blahut's theorem.

Suppose the field $GF(q^m)$ contains an element α of order N, i.e., an element such that $\alpha^N = 1$ but $\alpha^i \neq 1$ for $i = 1, 2, \cdots, N-1$. The discrete Fourier transform (DFT) of the $GF(q)$-ary sequence v^N is defined to be the $GF(q^m)$-ary sequence V^N, where

$$V_i = \sum_{j=0}^{N-1} v_j \alpha^{ij} \tag{23}$$

for $i = 0, 1, \cdots, N-1$.

Theorem 5.42 Over $GF(q)$, a field of characteristic P, a sequence v^N and its DFT are related by

$$V_j = \sum_{i=0}^{N-1} v_i \alpha^{ij}$$

$$v_i = \frac{1}{N}\sum_{j=0}^{N-1} V_j \alpha^{-ij} \tag{24}$$

where N is interpreted as an integer of the field, that is, modulo P, and (24) is called the inverse DFT.

Proof [Blah 83]: In any field

$$x^N - 1 = (x-1)(x^{N-1} + x^{N-2} + \cdots + x + 1)$$

By the definition of α, α^r is a zero of the left side for all r. Hence for all $r \neq 0 \bmod N$, α^r is a zero of the last term. But this is equivalent to

$$\sum_{j=0}^{N-1} \alpha^{ij} = 0, \qquad r \neq 0 \; (\bmod N)$$

whereas if $r = 0 \; (\bmod N)$

$$\sum_{j=0}^{N-1} \alpha^{rj} = N \qquad (\bmod\ P)$$

which is not zero if N is not a multiple of the field characteristic P. Combining these facts, we have

$$\sum_{j=0}^{N-1} \alpha^{-ij} \sum_{k=0}^{N-1} \alpha^{kj} v_k = \sum_{k=0}^{N-1} v_k \sum_{j=0}^{N-1} \alpha^{(k-i)j}$$
$$= (N \bmod P)v_i$$

Finally, $q^m - 1 = P^M - 1$ is a multiple of N, and consequently N is not a multiple of P. Hence $N \neq 0 \; (\bmod P)$. This proves the Theorem.

#

If one uses (23) to define V_i for all nonnegative integers i, one sees that

$$V_{i+N} = \sum_{j=0}^{N-1} v_j \alpha^{(i+N)j}$$
$$= \sum_{j=0}^{N-1} v_j \alpha^{ij}$$
$$= V_i$$

and hence that V^∞ is periodic with period N. Similarly, if one uses (24) to define v_i for all nonnegative integers j, one finds that v^∞ is periodic with period N. Thus, it is natural to consider that v^N and V^N are the first periods of the periodic sequences v^∞ and V^∞ with period N defined by (24) and (23) for all i and $j \geqslant 0$ respectively.

Theorem 5. 43   Let V^N be a sequence of elements of $GF(q^m)$ where N is a divisor of $q^m - 1$. Then the inverse Fourier transform v^N is a sequence of elements of $GF(q)$ if and only if the following equations are satisfied:

$$V_j^q = V_{((qj))} \qquad j = 0, 1, \cdots, N - 1$$

where $((qj)) = qj(mod\ N)$.

Proof [Blah 83]: By definition

$$V_j = \sum_{i=0}^{N-1} \alpha^{ij} v_i \qquad j = 0, 1, \cdots, N - 1$$

For a field of characteristic P, $(b + c)^{P^r} = b^{P^r} + c^{P^r}$ for any integer r. Further, if v_i is an element of $GF(q)$ for all i, then $v_i^q = v_i$. Consequently, combining these gives

$$V_j^q = \left(\sum_{i=0}^{N-1} \alpha^{ij} v_i \right)^q = \sum_{i=0}^{N-1} \alpha^{qij} v_i^q = \sum_{i=0}^{N-1} \alpha^{qij} v_i$$

$$= V_{((qj))}$$

Conversely, suppose that for all j, $V_j^q = V_{((qj))}$, then

$$\sum_{i=0}^{N-1} \alpha^{iqj} v_i^q = \sum_{i=0}^{N-1} \alpha^{iqj} v_i \qquad j = 0, 1, \cdots, N - 1$$

Let $k = qj$. Because q is relatively prime to $N = q^m - 1$, as j ranges over all values between 0 and $N - 1$, k also takes on all values between 0 and $N - 1$. Hence

$$\sum_{i=0}^{N-1} \alpha^{ik} v_i^q = \sum_{i=0}^{N-1} \alpha^{ik} v_i \qquad k = 0, 1, \cdots, N - 1$$

and by uniqueness of the Fourier transform, $v_i^q = v_i$ for all i. Thus, v_i is a zero of $x^q - x$ for all i, and such zeros are all elements of $GF(q)$.

$$\#$$

The following fundamental result, which is implicit in the work of Blahut [Blah 79], provides a link between linear complexity and DFT.

Blahut's Theorem   Let v^N be a sequence of length N over $GF(q)$, and let α be an element of order N in some extension field of $GF(q)$, and V^N the DFT of v^N, then

$$L(v^\infty) = W_H(V^N) \tag{25}$$

and

$$W_H(v^N) = L(V^\infty) \tag{26}$$

Proof: It follows from Theorem 5.3 that

$$L(v^\infty) = deg(f_v(x)) = N - deg(gcd(v^N(x), 1 - x^N))$$
$$= N - (\text{the number of roots of } v^N(x) \text{ which are}$$
$$N\text{-th roots of unity})$$
$$= W_H(V^N)$$

The remaining part of Blahut's theorem can be similarly proved. #

From the above proof, we can see that Blahut's theorem is simple. But it is useful: we now give the relationship between the weight complexity and fixed-complexity distance, and sphere complexity and variable-complexity distance of sequences by using Blahut's theorem.

Theorem 5. 44 Let s^∞ be a sequence of period N over $GF(q)$, and $GF(q)$ contain an element a of order N, S^N the DFT of s^N, then

$$WC_k(s^\infty) = FCD_k(S^\infty) \tag{27}$$
$$SC_k(s^\infty) = VCD_k(S^\infty) \tag{28}$$
$$FCD_k(s^\infty) = WC_k(S^\infty) \tag{29}$$
$$VCD_k(s^\infty) = SC_k(S^\infty) \tag{30}$$

Proof: Let t^∞ be any sequence of period N over $GF(q)$. For any sequence v^N over $GF(q)$, there must exist a sequence V^N over $GF(q)$ such that the inverse DFT of V^N is v^N, and vice versa. This is because $GF(q)$ contains an element of order N. Thus, it follows from Blahut's theorem that

$$WC_k(s^\infty) = \min_{W_H(t^N)=k} L(s^\infty + t^\infty)$$
$$= \min_{L((DFT(t^N))^\infty)=k} W_H((DFT(s^N + t^N))^N)$$
$$= \min_{L((T^N)^\infty)=k} W_H((DFT(s^N) + DFT(t^N))^N)$$
$$= \min_{L(T^\infty)=k} W_H(S^N + T^N)$$
$$= FCD_k(S^\infty)$$

This proves (27). The remaining parts can be similarly proved. #

Theorem 5. 45 Let s^∞ be a sequence with period N over $GF(q)$, and $GF(q^m)$ contain an element of order N , say α , S^N the $GF(q^m)$ -ary DFT of s^N , then

$$WC_k(s^\infty) \geqslant FCD_k(S^\infty) \tag{31}$$
$$SC_k(s^\infty) \geqslant VCD_k(S^\infty) \tag{32}$$

and

$$FCD_k(s^\infty) \geqslant WC_k(S^\infty) \tag{33}$$
$$VCD_k(s^\infty) \geqslant SC_k(S^\infty) \tag{34}$$

Proof: Noticing that not every sequence V^N over $GF(q^m)$ is DFT of a sequence v^N over $GF(q)$ if $m > 1$, and following the proof of Theorem 5. 44, we can prove the theorem.   #

Theorem 5. 46 Let s^∞ be a periodic sequence over $GF(q)$. Then

   a) If $FCD_k(s^\infty) = h$, then $WC_h(s^\infty) \leqslant k$
   b) If $WC_h(s^\infty) = k$, then $FCD_k(s^\infty) \leqslant h$

Proof: Trivial.   #

5. 7. 2   Bounds on the VCD of Binary Sequences with Period 2^n

In this section, we shall give some bounds on the variable-complexity distance of binary sequences with period 2^n . All the sequences discussed in this section are binary. Before developing the bounds, we would like to mention the relationship between the VCD of sequences and coding theory.

Let t^∞ be a sequence with period 2^n . Noticing that $t^\infty(x) = t^{2^n}(x)/(1 - x^{2^n}) = t^{2^n}(x)/(1 - x)^{2^n}$, we see that $L(t^\infty) \leqslant k$ iff $(1 + x)^{2^n}$ divides $t^{2^n}(x)$. Setting

$$C = \{c(x): c(x) = w(x)(1 + x)^{2^n - k}, deg(w(x)) \leqslant k - 1\}$$

then C is a $(2^n, k)$ -linear cyclic code with generating polynomial $(1 + x)^{2^n - k}$. Then it follows that

$$VCD_k(s^\infty) = \min_{c(x) \in C} W_H(s^{2^n}(x) + c(x))$$

This means that $VCD_k(s^\infty)$ is the minimum distance between s^{2^n} and the cyclic code C. Thus, the determination of the VCD of such a sequence is equivalent to the minimum distance decoding of a $(2^n, k)$-cyclic code.

Theorem 5.47 Let s^∞ be a binary sequence with period 2^n, then

$$VCD_{2^{n-1}}(s^\infty) = W_H(s^N) - 2r$$

where $N = 2^n$, and

$$r = \sum_{i=0}^{2^{n-1}-1} s_i \, s_{i+2^{n-1}}$$

Proof: Let $N = 2^n$, and

$$v^N(x) = s^N(x) + w(x)(1+x)^{2^{n-1}}$$

Noticing that

$$v_i = \begin{cases} s_i + w_i, & i = 0, 1, \cdots, 2^{n-1} - 1 \\ s_i + w_{i-2^{n-1}}, & i = 2^{n-1}, \cdots, 2^n - 1 \end{cases}$$

and $x \oplus y = x + y - 2xy$ for $x, y \in GF(2)$, we get

$$W_H(v^N) = \sum_{i=0}^{2^{n-1}-1} (s_i \oplus w_i) + \sum_{i=0}^{2^{n-1}-1} (s_{i+2^{n-1}} \oplus w_i)$$

$$= W_H(s^N) + 2 \sum_{i=0}^{2^{n-1}-1} [1 - (s_i + s_{i+2^{n-1}})] w_i$$

Let

$$A = \{i: s_i = s_{i+2^{n-1}} = 0\}$$
$$B = \{i: s_i = s_{i+2^{n-1}} = 1\}$$

then we get

$$W_H(v^N) = W_H(s^N) + 2 \sum_{i \in A} w_i - 2 \sum_{i \in B} w_i$$

By choosing $w_i = 0$ for $i \in A$, $w_i = 1$ for $i \in B$ we have

$$VCD_{2^{n-1}}(s^\infty) = W_H(s^N) - 2r. \qquad \#$$

Remark: The best 2^{n-1}-approximation of s^{2^n} (i. e., the sequence with linear complexity less than or equal to 2^{n-1} and with minimum Hamming distance to s^{2^n}) may not be unique. The approximations consist of sequences of those t^∞ such that $t^{2^n}(x) = w(x)(1+x)^{2^{n-1}}$ and $w_i = 0$ for $i \in A$, $w_i = 1$ for $i \in B$, otherwise $w_i = 0$ or 1. For example, let $N = 8 = 2^3$. $s^N = 10010001$, then $L(s^\infty) = 8$. Noticing $r = 1$, we get

$VCD_4(s^\infty) = 1$, and the best 2^{n-1}-approximation of s^∞ are those t^∞ with $t^N = *001*001$, where $*$ is 0 or 1.

Let s^∞ be a sequence with period $N = 2^n$. We now consider $VCD_{2^k}(s^\infty)$ for $k \leqslant n - 2$. Let $t^N(x) = w(x)(1 + x^2)$, where $deg(w(x)) \leqslant 2^n - 2^k - 1 = (2^{n-k} - 1)2^k - 1$, and

$$T^N = T_0 T_1 \cdots T_{2^{n-k}-1}$$
$$S^N = S_0 S_1 \cdots S_{2^{n-k}-1} \qquad (35)$$
$$W = W_0 W_1 \cdots W_{2^{n-k}-1}$$

where T_i, S_i and W_i are defined as

$$T_i = (t_{i2^k} t_{i2^k+1} \cdots t_{i2^k+2^k-1})$$
$$S_i = (s_{i2^k} s_{i2^k+1} \cdots s_{i2^k+2^k-1}) \qquad (36)$$
$$W_i = (w_{i2^k} w_{i2^k+1} \cdots w_{i2^k+2^k-1})$$

Then we have

$$T_i = \begin{cases} W_0 & , & i = 0 \\ W_i + W_{i+1} & , & 1 \leqslant i \leqslant 2^{n-k} - 2 \\ W_{2^{n-k}-2} & , & i = 2^{n-k} - 1 \end{cases}$$

Setting

$$W_i = \sum_{j=0}^{i} S_j (mod 2)$$

for $0 \leqslant i \leqslant 2^{n-k} - 2$, then we get

$$W_H(s^N + t^N) = W_H(\sum_{i=0}^{2^{n-k}-1} S_i (mod 2))$$

Thus, we have proved the following theorem:

Theorem 5. 48 Let s^∞ be a sequence with period 2^n , S_j as defined in (36) for each i with $0 \leqslant i \leqslant 2^{n-k} - 1$, then for $k \leqslant n - 2$,

$$VCD_{2^k}(s^\infty) \leqslant W_H(\sum_{j=0}^{2^{n-k}-1} S_j (mod 2)) \qquad \#$$

Example 3 Let $N = 32 = 2^5$, $k = 3$, and $s^N = 10010110000111010101101001011011$. It is easy to see that $L(s^\infty) = 32$. Noticing that $S_0 \oplus S_1 \oplus S_2 \oplus S_3 = 10001010$, we get $VCD_8(s^\infty) \leqslant 3$.

Remark : This example shows that although the number of 0s is nearly the same as that of 1s in s^N , and its linear complexity is equal to its period, its linear complexity is not stable.

In what follows in this section, we shall derive a general lower bound on $VCD_k(\cdot)$ for binary sequences with period 2^a . Assume s^∞ is a sequence with linear complexity $L(s^\infty)$, let t^∞ be a sequence with period 2^a , and $L(t^\infty) < L(s^\infty)$. The reduced rational forms of $s^\infty(x)$ and $t^\infty(x)$ can be written respectively as

$$s^\infty(x) = r_s(x)/(1+x)^{L(s^\infty)}$$
$$t^\infty(x) = r_t(x)/(1+x)^{L(t^\infty)}$$

where $r_s(1) \neq 0$, $r_t(1) \neq 0$. Then

$$(s^\infty + t^\infty)(x) = \frac{r_s(x) + r_t(x)(1+x)^{L(s^\infty)-L(t^\infty)}}{(1+x)^{L(s^\infty)}}$$

$$= v(x)/(1+x)^{L(s^\infty)}$$

Since $v(1) = r_s(1) \neq 0$, so $L(s^\infty + t^\infty) = L(s^\infty)$. On the other hand, we have already seen from the proof of Theorem 5.35 that $W_H(u^N) \geqslant \lceil Per(u^\infty)/L(u^\infty) \rceil$ for any periodic sequence u^∞ , so we have

$$
\begin{aligned}
VCD_k(s^\infty) &= \min_{0<L(t^\infty)\leqslant k} W_H(s^\infty + t^\infty) \\
&\geqslant \min_{0<L(t^\infty)\leqslant k} \lceil Per(s^\infty + t^\infty)/L(s^\infty + t^\infty) \rceil \\
&= \min_{0<L(t^\infty)\leqslant k} \lceil Per(s^\infty)/L(s^\infty) \rceil \\
&= \lceil Per(s^\infty)/L(s^\infty) \rceil
\end{aligned}
$$

for any k with $k < L(s^\infty)$. Thus, we have proved the following theorem:

Theorem 5.49 Let s^∞ be a sequence with period 2^a , and linear complexity $L(s^\infty)$. Then for any $k < L(s^\infty)$, we have

$$VCD_k(s^\infty) \geqslant \lceil 2^a/L(s^\infty) \rceil$$

Finally, we shall develop the relationships between the function $FCD_k(\cdot)$ and the co-correlation function $R_{s,t}(\cdot)$ as well as $VCD_k(\cdot)$ and $R_{s,t}(\cdot)$ of binary sequences. Let s^∞ and t^∞ be two binary sequences with period N , the co-correlation function of s^∞ and t^∞ is defined as

$$R_{s,t}(k) = \sum_{i=0}^{N-1} (-1)^{s_i + t_{i+k}}$$

Noticing that

$$
\begin{aligned}
R_{s,t}(0) &= \sum_{i=0}^{N-1} (-1)^{s_i + t_i} \\
&= \#\{i: s_i = t_i, \ 0 \leqslant i \leqslant N-1\} \\
&\quad - \#\{i: s_i \neq t_i, \ 0 \leqslant i \leqslant N-1\} \\
&= N - 2\#\{i: s_i \neq t_i, \ 0 \leqslant i \leqslant N-1\} \\
&= N - 2W_H(s^N + t^N)
\end{aligned}
$$

where $\#(\cdot)$ denotes the number of elements of the set $(\cdot)$, we get

$$W_H(s^N + t^N) = \frac{1}{2}(N - R_{s,t}(0))$$

Thus, by definition we have the following theorem:

Theorem 5. 50 Let s^∞ be a sequence with period N, then

$$FCD_k(s^\infty) = \frac{1}{2}N - \frac{1}{2}\max_{L(t^\infty)=k} R_{s,t}(0)$$

$$VCD_k(s^\infty) = \frac{1}{2}N - \frac{1}{2}\max_{0<L(t^\infty)\leqslant k} R_{s,t}(0)$$

6 The Period Stability of Sequences

The linear complexity of sequences has strong connections with their periods. For instance, the linear complexity of a sequence is equal to or less than its period. It is natural for one to think that the linear complexity of a sequence is not "stable" if its period is not "stable". By stable we mean intuitively that a few symbols changed in the corresponding places of each period do not make the period of the sequence decrease too much. For example, let us consider the binary sequence:

$$s^\infty = (110010111001011100100)^\infty$$

It is easy to check that s^∞ has period 21, and its linear complexity is equal to or greater than 18. But one bit of change at the last bit of each period makes the period of s^∞ jump to 7, and the linear complexity to 3. Consequently, to make the linear complexity of a sequence stable, it is necessary to ensure a good period stability.

Before analyzing the period stability of sequences, it is imperative to prove some basic results about the orders of polynomials and periods of sequences. Section 6.1 fulfills this requirement. Section 6.2 first gives, from the viewpoint of stream ciphers, two measure indexes for the stability of periods, i.e., weight period and sphere period, then develops the relationships between weight period and weight complexity as well as sphere period and sphere complexity. It is also natural and intuitive for one to see that the weight period of sequences should have some links with the autocorrelation function of sequences. Section 6.3 investigates the links. To analyze the period stability, we need bounds on the weight period $WP_k(s^\infty)$. Sections 6.4 and 6.5 are devoted to the development of some bounds. Furthermore, a definite relationship beween the weight period and weight complexity as well as sphere period and sphere complexity of binary sequences with period 2^n is developed in Section 6. 5, in Theorem 6. 19.

6. 1 General Results about Orders of Polynomials and Periods of Sequences

Before investigating the stability of periods, it is necessary to make some notations clear and to prove some results about the order of polynomials and the period of sequences. By period we refer to the minimal period of sequences in this chapter.

Let $f \in GF(q)[x]$ be a polynomial of degree $m \geqslant 1$ with $f(0) \neq 0$. The residue class ring $GF(q)[x]/(f)$ contains $q^m - 1$ nonzero residue classes. The q^m residue classes $x^j + (f)$, $j = 0, 1, \cdots, q^m - 1$, are all nonzero, and so there exist integer r and s with $0 \leqslant r < s \leqslant q^m - 1$ such that $x^s = x^r \bmod f(x)$. Since x and $f(x)$ are relatively prime, it follows that $x^{s-r} = 1 \bmod (f)$, that is, $f(x)$ divides $x^{s-r} - 1$, and $0 < s - r \leqslant q^m - 1$. Thus there exists a positive integer $e \leqslant q^m - 1$ such that $f(x)$ divides $x^e - 1$ for any polynomial $f(x)$ over $GF(q)$ with $f(0) \neq 0$. Because of this fact the following definition is rational.

Definition 6. 1 Let $f \in G(q)[x]$ be a nonzero polynomial. If $f(0) \neq 0$, then the least positive integer e for which $f(x)$ divides $x^e - 1$ is called the order of f and denoted by $ord(f) = ord(f(x))$. If $f(0) = 0$, then $f(x) = x^m g(x)$, where m, a positive integer, and $g \in GF(q)[x]$ with $g(0) \neq 0$ are uniquely determined, $ord(f)$ is then defined to be $ord(g)$.

The order of a polynomial $f(x)$ is also called the exponent or period of $f(x)$. Let $f(x) \in GF(q)[x]$ be an irreducible polynomial over $GF(q)$ of degree m and with $f(0) \neq 0$. It is easy to see that $ord(f(x))$ is equal to the order of any root of $f(x)$ in the multiplicative group $GF(q^m)^*$, where $GF(q^m)^* = GF(q^m) - \{0\}$. It follows from this fact that $ord(f(x))$ divides $q^m - 1$ if $f(x) \in GF(q)[x]$ is irreducible over $GF(q)$.

Lemma 6. 2 Let c be a positive integer. Then the polynomial $f(x) \in GF(q)[x]$ with $f(0) \neq 0$ divides $x^c - 1$ iff $ord(f(x))$ divides c.

Proof: If $e = ord(f(x))$ divides c, then $f(x)$ divides $x^e - 1$ and $x^e - 1$ divides $x^c - 1$, so that $f(x)$ divides $x^c - 1$. Conversely, if $f(x)$ divides $x^c - 1$ we have $c \geqslant e$, so that we can write $c = me + r$ with m being a positive integer, and $0 \leqslant r < e$. Since $x^c - 1 = (x^{me} - 1)x^r + (x^r - 1)$, it follows that $f(x)$ divides $x^r - 1$, which is only possible for $r = 0$. Therefore, e divides c. #

Since any polynomial of positive degree can be written as a product of irreducible polynomials, the computation of orders of polynomials can be achieved if one knows how to determine the order of a power of an irreducible polynomial and the order of the product of pairwise relatively prime polynomials. Since powers of x are factored out in advance when determining the order of a polynomial, we need not consider the power of the irreducible polynomial $g(x)$ with $g(0) = 0$.

Theorem 6. 3 Let $g \in GF(q)[x]$ be irreducible over $GF(q)$ with $g(0) \neq 0$ and $ord(g) = e$, and let $f(x) = g^b(x)$ with a positive integer b. Let t be the smallest integer with $p^t \geqslant b$, where p is the characteristic of $GF(q)$. Then ord $(f(x)) = ep^t$.

Proof[Lidi 85]: Setting $c = ord(f(x))$ and noting that the divisibility of $x^c - 1$ by $f(x)$ implies the divisibility of $x^c - 1$ by $g(x)$, we obtain that e divides c by Lemma 6. 2. Furthermore, $g(x)$ divides $x^e - 1$, therefore $f(x)$ divides $(x^e - 1)^b$ and it divides $(x^e - 1)^{p^t} = x^{ep^t} - 1$. Thus according to Lemma 6. 2, c divides ep^t. It follows from what we have shown so far that c is of the form $c = ep^u$ with $0 \leqslant u \leqslant t$. We note now that $x^e - 1$ has only simple roots, since e is not a multiple of p. Therefore, all the roots of $x^{ep^u} - 1 = (x^e - 1)^{p^u}$ have multiplicity p^u. But $g(x)^b$ divides $x^{ep^u} - 1$, whence $p^u \geqslant b$ by computing multiplicities of roots, and so $u \geqslant t$, Thus we get $u = t$ and $c = ep^t$.

Theorem 6. 4 Let $g_1, g_2, \cdots, g_k$ be pairwise relatively prime nonzero polynomials over $GF(q)$, and let $f = g_1 g_2 \cdots g_k$. Then $ord(f(x))$ is equal to the least common multiple of $ord(g_1(x)), \cdots, ord(g_k(x))$.

Proof [Lidi 85]: It is easily seen that it suffices to consider the case where $g_i(0) \neq 0$ *for* $1 \leqslant i \leqslant k$, set $e = ord(f(x))$ and $e_i = ord(g_i(x))$ for $1 \leqslant i \leqslant k$, and let $c = lcm\ (e_1, e_2, \cdots, e_k)$. Then each $g_i(x)$, $1 \leqslant i \leqslant k$, divides $x^{e_i} - 1$, and so $g_i(x)$ divides $x^c - 1$. Because of the pairwise relative primality of the polynomials $g_1(x)$, $g_2(x)$, $\cdots$, $g_k(x)$, we obtain that $f(x)$ divides $x^c - 1$. An application of Lemma 6. 2 shows that e divides c . On the other hand, $f(x)$ divides $x^e - 1$, and so each $g_i(x)$, $1 \leqslant i \leqslant k$, divides $x^e - 1$. Again by Lemma 6. 2, it follows that each e_i, $1 \leqslant i \leqslant k$, divides e , therefore c divides e . Thus we conclude that $e = c$. #

Example 1 Let us compute the order of $f(x) = x^{10} + x^9 + x^3 + x^2 + 1$ $\in GF(2)[x]$. The canonical factorization of $f(x)$ over $GF(2)$ is given by $f(x) = (x^2 + x + 1)^3 (x^4 + x + 1)$. Since $ord(x^2 + x + 1) = 3$, we get $ord((x^2 + x + 1)^3) = 12$ by Theorem 6. 3. Furthermore, $ord(x^4 + x + 1) = 15$, and so Theorem 6. 4 implies that $ord(f(x))$ is equal to the least common multiple of 12 and 15, that is, $ord(f(x)) = 60$.

Theorem 6. 5 Let $GF(q)$ be a finite field of characteristic p , and let $f(x) \in GF(q)[x]$ be a polynomial of positive degree and with $f(0) \neq 0$. Let $f(x) = af_1^{b_1}(x) \cdots f_k^{b_k}(x)$, where $a \in GF(q)$, b_1, b_2, $\cdots$, b_k are positive integers, and $f_1(x)$, $f_2(x), \cdots$, $f_k(x)$ are distinct monic irreducible polynomials in $GF(q)[x]$, be the canonical factorization of $f(x)$ in $GF(q)[x]$. Then $ord(f(x)) = ep^t$, where e is the common multiple of $ord(f_1(x))$, $\cdots$, $ord(f_k(x))$, and t is the smallest integer with $p^t \geqslant max(b_1, b_2, \cdots, b_k)$.

Proof: Based on Theorems 6. 3 and 6. 4, it is trival to see that Theorem 6. 5 is valid. #

In what precedes, we have present some important results about the order of polynomials, which will be useful in the following sections. In what follows in this section, we shall provide some basic results about the period of sequences. By period we refer to the minimal period of a sequence in this chapter.

Theorem 6. 6 Let $f_s(x)$ be the minimal polynomial of the periodic sequence s^∞ over $GF(q)$. Then $Per(s^\infty) = ord(f_s(x))$, where here and hereafter $Per(s^\infty)$ denotes the period of s^∞.

Proof: Assume that the reduced rational form of $s^\infty(x)$ is:

$$s^\infty(x) = r_s(x)/f_s(x), \qquad gcd(r_s(x), f_s(x)) = 1$$

Let N be the period of $s^\infty(x)$, then $f_s(x)$ divides $1 - x^N$. By Lemma 6. 2 we get that $ord(f_s(x))$ divides N. Suppose $ord(f_s(x)) < N$, setting

$$g(x) = (1 - x^{ord(f_s(x))})/f_s(x)$$

we have

$$s^\infty(x) = r_s(x)g(x)/f_s(x)g(x)$$
$$= r_s(x)g(x)/(1 - x)^{ord(f_s(x))}$$

where $deg(r_s(x)g(x)) < ord(f_s(x))$. Therefore, the period of $s^\infty(x)$ is equal to or less than $ord(f_s(x)) < N$. This is contrary to the minimality of N. Hence $ord(f_s(x)) = N$. #

Based on Theorem 5. 5 and Theorem 6. 6, one can arrive at the following conclusion.

Theorem 6. 7 Let s^∞ and t^∞ be two sequences over $GF(q)$. If the reduced rational forms of $s^\infty(x)$ and $t^\infty(x)$ are $s^\infty(x) = r_s(x)/f_s(x)$ and $t^\infty(x) = r_t(x)/f_t(x)$ respectively, then

$$Per(s^\infty + t^\infty) = ord[f_s(x)f_t(x)/gcd(f_s(x)f_t(x),$$
$$r_s(x)f_t(x) + f_s(x)r_t(x))] \qquad \#$$

Theorem 6. 8 Let s^∞ and t^∞ be two sequences with reduced rational forms of $s^\infty(x)$ and $t^\infty(x)$ as $s^\infty(x) = r_s(x)/f_s(x)$ and $t^\infty(x) = r_t(x)/f_t(x)$. Then the sequence $s^\infty * t^\infty$ (the convolution sequence of s^∞ and t^∞) has minimal polynomial

$$f_s(x)f_t(x)/gcd(f_s(x)f_t(x), r_s(x)r_t(x))$$

and period

$$ord[f_s(x)f_t(x)/gcd(f_s(x)f_t(x), r_s(x)r_t(x))]$$

Proof: By assumptions we get

$$(s^\infty * t^\infty)(x) = r_s(x)r_t(x)/f_s(x)f_t(x)$$

Thus, the minimal polynomial of $s^\infty * t^\infty$ is

$$f_s(x)f_t(x)/gcd(f_s(x)f_t(x), r_s(x)r_t(x))$$

It follows from this fact and Theorem 6.6 that the period of $s^\infty * t^\infty$ is

$$ord[f_s(x)f_t(x)/gcd(f_s(x)f_t(x), r_s(x)r_t(x))] \quad \#$$

6.2 Measure Indexes for the Stability of Period and Their Relationships with Weight Complexity and Sphere Complexity

From the viewpoint of stream ciphers, we define, for the purpose of measuring the stability of stream ciphers, the weight period or sphere surface period and sphere period respectively as follows.

Definition 6.9 Let s^∞ be a sequence of period N. The weight and sphere period of s^∞ are respectively defined as

$$WP_k(s^\infty) = \min_{W_H(t^N)=k} Per(s^\infty + t^\infty)$$

and

$$SP_k(s^\infty) = \min_{0 < W_H(t^N) \leqslant k} Per(s^\infty + t^\infty)$$

where t^∞ is a sequence of period N.

The basic properties of $WP_k(s^\infty)$ and $SP_k(s^\infty)$ presented in the following theorem are trival. It is easy to check these properties.

Theorem 6.10 The weight period and sphere period have the following basic properties:

1) $WP_k(s^\infty) | Por(s^\infty)$;
2) $SP_k(s^\infty) | Per(s^\infty)$;
3) $WC_k(s^\infty) \leqslant WP_k(s^\infty)$;
4) $SC_k(s^\infty) \leqslant SP_k(s^\infty)$.

Based on Definition 6.9 and Theorem 6.7, one can easily derive the following conclusion.

Theorem 6. 11 Let s^∞ be a sequence of period N where the reduced rational form of $s^\infty(x)$ is $r_s(x)/f_s(x)$. Then

$$WP_k(s^\infty) = \min_{W_H(t^N)=k} ord[f_s(x)f_t(x)/ged(f_s(x)f_t(x),$$

$$r_s(x)f_t(x) + f_s(x)r_t(x))]$$

$$SP_k(s^\infty) = \min_{1<i\leqslant k} \min_{W_H(t^N)=i} ord[f_s(x)f_t(x)/$$

$$gcd(f_s(x)f_t(x), \; r_s(x)f_t(x) + f_s(x)r_t(x))]$$

where t^∞ is periodic with period N and $r_t(x)/f_t(x)$ is the reduced rational form of $t^\infty(x)$. $\qquad\qquad$ #

We now investigate the relationships between weight complexity and weight period as well as sphere complexity and sphere period. The relationships make some of their results interchangable.

Theorem 6. 12 Let s^∞ be a periodic sequence over $GF(q)$. Then

$$\lceil log_q WP_k(s^\infty) \rceil \leqslant WC_k(s^\infty) \tag{1}$$

$$\lceil log_q SP_k(s^\infty) \rceil \leqslant SC_k(s^\infty) \tag{2}$$

Proof: We first show that

$$\lceil log_q ord(f(x)) \rceil \leqslant deg(f(x)) \tag{3}$$

for any $f(x) \in GF(q)[x]$. We have already shown, at the beginning of Section 6. 1, that $ord(f(x))$ divides $q^m - 1$ if $f(x) \in GF(q)[x]$ is an irreducible polynomial of degree m over $GF(q)$. Thus inequality (3) is valid in the case that $f(x)$ is irreducible. For the case $f(x) = g^b(x)$, where $g(x)$ is irreducible and $g(0) \neq 0$, suppose $ord(g(x)) = e$ and t is the smallest integer such that $p^t \geqslant b$, where p is the characteristic of the field $GF(q)$. From the conclusion above we get

$$\lceil log_q ord(g(x)) \rceil \leqslant deg(g(x)).$$

It follows from Theorem 6. 3 that $ord(f(x)) = ep^t$. Therefore, we have got

$$\lceil log_q ord(f(x)) \rceil = \lceil log_q ep^t \rceil$$

$$= \lceil log_q ord(g(x))p^t \rceil$$

$$= \lceil log_q ord(g(x)) + log_q p^t \rceil$$

$$= \lceil log_q ord(g(x)) + \lceil log_p b \rceil log_q p \rceil$$
$$\leqslant \lceil log_q ord(g(x)) \rceil + \lceil log_p b \rceil$$
$$\leqslant deg(g(x)) + \lceil log_p b \rceil$$
$$\leqslant b \cdot deg(g(x))$$
$$= deg(f(x))$$

for $b > 1$ and $deg(g(x)) \geqslant 1$. Thus, the inequality (1) holds for the case $f(x) = g^b(x)$. For any $f(x) \in GF(q)[x]$, let

$$f(x) = f_1^{b_1}(x) f_2^{b_2}(x) \cdots f_k^{b_k}(x)$$

where $b_1, b_2, \cdots, b_k$ are positive integers, and $f_1(x), f_2(x), \cdots, f_k(x)$ are distinct monic irreducible polynomial over $GF(q)$, be the canonical factorization of $f(x)$ in $GF(q)[x]$. Setting

$$g_i(x) = f_i^{b_i}(x), \qquad i = 1, 2, \cdots, k$$

we get $f(x) = g_1(x) g_2(x) \cdots g_k(x)$, where $g_1(x), \cdots, g_k(x)$ are pairwise relatively prime polynomials over $GF(q)$. By the above conclusion we get

$$\lceil log_q ord(g(x)) \rceil \leqslant deg(g_i(x)), \qquad i = 1, 2, \cdots, k.$$

Thus, it follows from Theorem 6. 4 that

$$ord(f(x)) = lcm \lceil ord(g_1(x)), ord(g_2(x)), \cdots, ord(g_k(x)) \rceil$$
$$\leqslant ord(g_1(x)) ord(g_2(x)) \cdots ord(g_k(x))$$

Hence, we have

$$\lceil log_q ord(f(x)) \rceil \leqslant \lceil log_q ord(g_1(x)) ord(g_2(x)) \cdots ord(g_k(x)) \rceil$$
$$\leqslant \sum_{i=1}^{k} \lceil log_q ord(g_i(x)) \rceil$$
$$\leqslant \sum_{i=1}^{k} deg(g_i(x))$$
$$= deg(f(x))$$

So far we have completed the proof of the inequality (1). Noticing Theorem 6. 11 and Theorem 5. 6, we get

$$WC_k(s^\infty) = \min_{W_H(t^N)=k} deg[f_s(x) f_t(x)/gcd(f_s(x) f_t(x),$$
$$r_s(x) f_t(x) + f_s(x) r_t(x))]$$
$$\geqslant \min_{W_H(t^N)=k} \lceil log_q ord[f_s(x) f_t(x)/gcd(f_s(x) f_t(x),$$

$$r_s(x)f_t(x) + f_s(x)r_t(x))\rceil$$
$$= \lceil log_q WP_k(s^\infty)\rceil$$

Similarly, we can prove that

$$SC_k(s^\infty) \geqslant \lceil log_q SP_k(s^\infty)\rceil \ \#$$

Remarks : a) The equality in (3) holds iff $f(x)$ is primitive, i. e. , $f(x)$ is monic of degree m with $ord(f(x)) = q^m - 1$ and $f(0) \neq 0$.

b) The equality in (1) and (2) can be achieved. For example, let $s^\infty = (01)^\infty, k = 1$. Then it is easy to check that $WC_1(s^\infty) = \lceil log_2 WP_1(s^\infty)\rceil$. A sequence s^∞ with period N such that the equality in (1) holds, for a positive integer k, is called a (N, k)-maximum-length sequence.

6. 3 The Weight Period and the Autocorrelation Function of Binary Sequences

It can be seen intuitively that there should exist some connections between the weight period and the autocorrelation function of sequences. Actually, there do exist some. We now develop some of them for binary sequences. Let s^∞ be a binary sequence of period N. The autocorrelation function of s^∞ is defined as

$$C_s(k) = (\sum_{i=1}^{N-1} (-1)^{s_i+s_{i+k}})/N \tag{4}$$

Theorem 6. 13 Let s^∞ be a binary sequence with an even period N. Then

$$WP_{N[1-c_s(N/2)]/4}(s^\infty) \leqslant N/2 \tag{5}$$

Proof: Let $k = N/2$. Notice that

$$C_s(k) = (\sum_{i=0}^{N-1} (-1)^{s_i+s_{i+N}})/N$$
$$= [N - 4\#\{i:s_i \neq s_{i+k}, 0 \leqslant i \leqslant k-1\}]/N,$$

we get

$$\#\{i:s_i \neq s_{i+k}, 0 \leqslant i \leqslant k-1\} = N[1 - C_s(N/2)]/4$$

Thus, by the definition of $WP_k(s^\infty)$ it follows that

$$WP_{N[1-C_s(N/2)]/4}(s^\infty) \leqslant N/2 \quad \#$$

Remark : The equality in (5) can be achieved. For example, let $s^\infty =$ 1110010011100100 $\cdots$, $N = 8$. Then $C_s(N/2) = C_s(2) = 0$ and $N(1 - C_s(N/2))/4 = 2$. It is easy to verify that $WP_2(s^\infty) = 4 = N/2$.

Theorem 6. 14 Let s^∞ be a binary sequence of period N, and M a positive integer with M dividing N and $M < N$. If $WP_k(s^\infty) = M$, and $k < M/4$, then

$$C_s(iM) \geqslant 1 - 4k/N$$

for any positive integer i.

Proof: Let w^∞ be a binary sequence of period N with $W_H(w^N) = k$, and $Per(w^\infty + s^\infty) = M$. Setting $t^\infty = w^\infty + s^\infty$, we get $t_i = t_{i+jM}$ for any i. Therefore, it follows that

$$
\begin{aligned}
C_s(jM) &= (\sum_{i=0}^{N-1} (-1)^{s_i + s_{i+jM}})/N \\
&= (\sum_{i=0}^{N-1} (-1)^{t_i + w_i + t_{i+jM} + w_{i+jM}})/N \\
&= (\sum_{i=0}^{N-1} (-1)^{w_i + w_{i+jM}})/N \\
&= (N - 2\#\{i : w_i \neq w_{i+jM},\ 0 \leqslant i \leqslant N - 1\})/N \\
&\geqslant (N - 4k)/N \\
&= 1 - 4k/N \qquad\qquad\qquad\qquad\qquad \#
\end{aligned}
$$

6. 4 Bounds on the Weight Complexity $WP_k(s^\infty)$ for $1 \leqslant k \leqslant 2$

In this section, in order to analyze the stability of periods of sequences, we shall give some results about $WP_1(s^\infty)$ and $WP_2(s^\infty)$.

Theorem 6. 15 Let s^∞ be a sequence of period N, $f_s(x)$ the minimal polynomial of s^∞ and $g(x) = (1 - x^N)/f_s(x)$. Let

$$f_s(x) = f_1^{b_1}(x)f_2^{b_2}(x)\cdots f_k^{b_k}(x)$$

be the canonical factorization of $f_s(x)$ in $GF(q)[x]$, and e the largest positive integer such that $f^e(x)$ divides $1-x^N$. Then

1) If $e \leqslant 2$, then $WP_1(s^\infty)=N$

2) if $e=1$, then

$$WP_1(s^\infty) \geqslant lcm\{ord\ f_1(x),\ ord\ f_2(x),\ \cdots,\ ord\ f_k(x)\}$$
$$\times \lceil log_P max(b_1,b_2,\cdots,b_k)\rceil$$

where P is the characteristic of $GF(q)$.

Proof: Let t^∞ be a sequence of period N over $GF(q)$ with $W_H(t^N)=1$. Then $t^\infty(x)=x^i/(1-x^N)$, $o\leqslant i\leqslant N-1$, and $f_t(x)=1-x^N$, $r_t(x)=x^i$. Thus, it follows from Theorem 6.7 that

$$WP_1(s^\infty) = \min_{0\leqslant i\leqslant N-1}\ ord\ [(1-x^N)/gcd(1-x^N,x^i+$$
$$r_s(x)(1-x^N)/f_s(x))]$$

If $e\geqslant 2$, it is easy to see that

$$h_i(x) = gcd\ (1-x^N,\ x^i+r_s(x)(1-x^N)/f_s(x)) = 1$$

for each i with $0\leqslant i\leqslant N-1$. Therefore, we get $WP_1(s^\infty)=N$. If $e=1$, then $h_i(x)$ is equal to 1 or $f_s(x)$. Let us assume that $h_i(x)=f_s(x)$ for some i with $0\leqslant i\leqslant N-1$, then it follows from Theorem 6.5 that

$$WP_1(s^\infty) = min[N,\ lcm(ord(f_1(x)),\cdots,ord\ (f_k(x)))$$
$$\times \lceil log_p max(b_1,b_2,\cdots,b_k)\rceil]$$
$$\geqslant lcm(ord(f_1(x)),\cdots,ord(f_k(x)))\lceil log_P max(b_1,b_2,\cdots,b_k)\rceil \quad \#$$

Theorem 6.16 Let s^∞ be a binary ML-sequence of period 2^m-1. If $f_s(x)$ is not equal to its reciprocal polynomial, then $WP_1(s^\infty)=2^m-1$.

Proof: Since $f_s(x)$ is primitive, $ord(f_s(x))=2^m-1$. It is easy to see that the order of the reciprocal polynomial $f_s^*(x)$ is equal to $ord(f_s(x))=2^m-1$. Therefore, it follows from $f_s^*(x)\neq f_s(x)$ that $f_s^*(x)f_s(x)$ divides $Q^{(2^m-1)}(x)$, where $Q^{(2^m-1)}(x)$ is the cyclotomic polynomial defined as

$$Q^{(2^m-1)}(x) = \prod_{w\in W}(x-w)$$

where W is the set of field elements with order 2^m-1. Thus, $g(x)=$

$(1+x^{2^m-1})/f_s(x)$ contains a primitive polynomial of degree m. Hence, by Theorem 6.15 we obtain $WP_1(s^\infty)=2^m-1$.   #

Example 2   Let $s^\infty=101101\cdots$, be a binary ML-sequence of period 3. Then the minimal polynomial of s^∞ is x^2+x+1. It is easy to verify that $WP_1(s^\infty)=1<2^2-1=3$. This is due to the fact that the reciprocal polynomial of x^2+x+1 is itself.

Example 3   Noticing that

$$x^{15}+1 = (x+1)(x^2+x+1)(x^4+x^3+x^2+x+1)$$
$$\times (x^4+x^3+1)(x^4+x+1)$$

over $GF(2)$, we have $WP_1(s^\infty)=15$ for any binary ML-sequence of period 15, where

$$Q^{(15)}(x) = (x^4+x^3+1)(x^4+x+1)$$

Theorem 6.17   Let s^∞ be a binary ML-sequence of period 2^m-1. If the reciprocal polynomial of $f_s(x)$ is not itself, then

$$WP_2(s^\infty) = 2^m - 1$$

Proof: Let t^∞ be a binary sequence of period 2^m-1 with $W_H(t^N)=2$, where $N=2^m-1$. Suppose that $t_i=t_j=1$ and $t_k=0$ for all $k\neq i$ and j, then $t^\infty(x)=(x^i+x^j)/(1+x^N)$. Without loss of generality, assume $i<j$ and set $M=gcd(j-i,N)$, then we get

$$f_t(x) = (1+x^N)/(1+x^M),$$
$$r_t(x) = (x^i+x^j)/(1+x^M).$$

It follows from Theorem 6.11 that

$$WP_2(s^\infty) = \min_{0\leqslant i<j\leqslant N-1} ord[f_s(x)f_t(x)/gcd(f_s(x)f_t(x),$$
$$r_s(x)f_t(x) + f_s(x)r_t(x))]$$

$$= \min_{0\leqslant i<j\leqslant N-1} ord[(1+x^N)/gcd(1+x^N,$$
$$x^i+x^j+r_s(x)(1+x^N)/f_s(x))]$$

Since $0<j-i\leqslant N-1$, and the reciprocal polynomial of $f_s(x)$ is primitive, it does not divide x^i+x^j. Let

$$U_{i,j}(x) = gcd[1+x^N, x^i+x^j+r_s(x)(1+x^N)/f_s(x)]$$

then $f_s^*(x)$ does not divide $U_{i,j}(x)$. Therefore $(1+x^N)/U_{i,j}(x)$ contains the factor $f_s^*(x)$ for each pair (i,j) with $i<j$. On the other hand, since $N=(2^m-1)$ is odd, it follows from Lemma 5.24 that $1+x^N$ has no repeated factor. Thus it follows that

$$WP_2(s^\infty) = \min_{0\leqslant i<j\leqslant N-1} ord[(1+x^N)/U_{i,j}(x)]$$
$$\geqslant \min_{0\leqslant i<j\leqslant N-1} ord(f_s^*(x)) = N$$

It is also apparent that $WP_2(s^\infty)<N$, hence $WP_2(s^\infty)=N=2^m-1$.

$$\#$$

We now investigate the weight period $WP_1(s^\infty)$ for the output sequences of the stop-and-go generator depicted in Fig 5.2. Let $f_1(x)$ and $f_2(x)$ be the primitive feedback polynomials of degree m. Lemma 5.32 tells us that the output sequence has minimal polynomial $f_2(x^{2^m-1})$ and period $(2^m-1)^2$. Let $M=2^m-1$. Noticing that x^M+1 has no repeated factor, we assume that

$$x^M + 1 = g_1(x)g_2(x)\cdots g_k(x)$$

where $g_1(x)$, $g_2(x)$, $\cdots$, $g_k(x)$ are irreducible polynomials over $GF(2)$, is the canonical factorization of x^M+1 over $GF(2)$. Then

$$x^{M^2} + 1 = g_1(x^M)g_2(x^M)\cdots g_k(x^M)$$

It follows that, if $f_2(x^M)$ divide $x^{M^2}+1$, $f_2^*(x^M)$ divides $x^{M^2}+1$. Since $ord(f_2(x^M))=M^2$, $ord(f_s^*(x^M))=M^2$. Thus we arrive at the following conclusion.

Theorem 6.18　Let $f_1(x)$ and $f_2(x)$ be the primitive feedback polynomials of SR1 and SR2 respectively with degree m. If $f_2(x)$ is not equal to its reciprocal polynomial, then

$$WP_1(s^\infty) = (2^m - 1)^2 \qquad\qquad \#$$

6. 5 The Period Stability of Binary Sequences with Period 2^n

In this section, we shall show that the connections between $WP_k(s^\infty)$ and $WC_k(s^\infty)$ are especially strong for binary sequences with period 2^n. The main purpose of this section is to develop bounds on $WP_k(s^\infty)$ for binary sequences with period 2^n. Let us stipulate that all the sequences are binary in this section.

Theorem 6. 19 Let s^∞ be a sequence of period 2^n, then

$$WP_k(s^\infty) = 2^{\lceil log_2 WC_k(s^\infty) \rceil}$$

$$SP_k(s^\infty) = 2^{\lceil log_2 SC_k(s^\infty) \rceil}$$

To prove Theorem 6. 19, we need to prove the following lemma first. The lemma shows that the link between the period and linear complexity of binary sequences with period 2^n is very close.

Lemma 6. 20 Let s^∞ be a sequence with period 2^n , then

$$Per(s^\infty) = 2^{\lceil log_2 L(s^\infty) \rceil}$$

Proof: Since $Per(s^\infty) = 2^n$, the minimal polynomial $f_s(x) = (1 + x)^{L(s^\infty)}$. Note that if $l = 2^k \cdot p$ with p an odd positive integer, then

$$1 + x^l = (1 + x)^{2^k} g(x)$$

where $g(x) = (x^{2^k(p-1)} + x^{2^k(p-2)} + \cdots + x^{2^k} + 1)$ with $g(1) = 1$ over $GF(2)$. It follows that

$$ord[(1 + x)^{L(s^\infty)}] = 2^{\lceil log_2 L(s^\infty) \rceil}$$

This proves the lemma.   #

Corollary 6. 21 If s^∞ is a sequence with period 2^n , then

$$2^{n-1} + 1 \leqslant L(s^\infty) \leqslant 2^n$$

Proof of Theorem 6. 19: It follows from Lemma 6. 20 that

$$WP_k(s^\infty) = \min_{W_H(t^N)=k} Per(s^\infty + t^\infty)$$

$$= \min_{W_H(t^N)=k} 2^{\lceil log_2 L(s^\infty + t^\infty) \rceil}$$

$$= 2^{\lceil log_2 \min_{W_H(t^N)=k} L(s^\infty + t^\infty) \rceil}$$

$$= 2^{\lceil log_2 WC_k(s^\infty) \rceil}$$

Similarly, we can prove the remaining part of Theorem 6. 19.   #

Theorem 6. 19 gives a definite relationship between the weight period $WP_k(s^\infty)$ and weight complexity $WC_k(s^\infty)$ for binary sequences with period 2^n. It plays the role of a bridge between $WP_k(s^\infty)$ and $WC_k(s^\infty)$, and allow us to get results about one by making use of those about the other. The following theorems can illustrate this.

Theorem 6. 22 Let s^∞ be a sequence of period 2^n. If $L(s^\infty) < 2^n$, then $WP_1(s^\infty) = 2^n$.

Proof: It follows directly from Theorem 5. 7 and Theorem 6. 19.   #

Theorem 6. 23 Let s^∞ be a sequence with period 2^n. If $L(s^\infty) = 2^n$ and $r_s(x) = 1 + x^j + x^{2^{n-1}}$, where $j = 2^e j_1$, with j_1 an odd number, $0 < j \leqslant 2^n - 1$ and $j \neq 2^{n-1}$, then

$$WP_1(s^\infty) = 2^{\lceil log_2(2^n - 2^e) \rceil}$$

Proof: It follows directly from Theorem 6. 19 and Theorem 5. 11.

Theorem 6. 24 Let s^∞ be periodic with period 2^n. If $L(s^\infty) = 2^n$ and

$$r_s(x) = 1 + x^{2^{t_1}} + \cdots + x^{2^{t_v}} \qquad 4 \leqslant v \leqslant n - 1,$$

where v is even and $t_1 < t_2 < \cdots < t_v$, then

$$WP_1(s^\infty) = 2^{t_1 + \lceil log_2(2^{n-t_1} - 2^{t_2-t_1} - 1) \rceil}$$

Proof: It follows directly from Theorem 6. 19 and Theorem 5. 12.   #

Theorem 6. 25 Let s^∞ be periodic with period 2^n. If $L(s^\infty) = 2^n$ and

$$r_s(x) = 1 + x^{i_1} + \cdots + x^{i_m}$$

where m is even and $i_v = 2^{e_v} p_v$ with p_v an odd number, $1 \leqslant v \leqslant m$ and $e_1 < e_2 < \cdots < e_m$, then

$$WP_1(s^\infty) \geqslant 2^{e_1 + \lceil log_2(2^{n-e_1} - 2^{e_2-e_1} - 1) \rceil}$$

Proof: It follows directly from Theorem 6. 19 and Theorem 5. 14.   #

Theorem 6. 26 Let s^∞ be periodic with period 2^n. If $L(s^\infty) = 2^n$, then $WP_2(s^\infty) = 2^n$.

Proof: It follows directly from Theorem 5. 17 and Theorem 6. 19.   #

Theorem 6. 27 Let s^∞ be periodic with period 2^n.

   a) If $L(s^\infty) < 2^{n-1}$, then $WP_2(s^\infty) = 2^{n-1}$

   b) If $L(s^\infty) > 2^{n-1}$, and $L(s^\infty) \neq 2^n - 2^m$ for every m with
    $0 \leqslant m \leqslant n - 1$, then

$$WP_2(s^\infty) = 2^{\lceil \log_2 L(s^\infty) \rceil}$$

Proof: It follows directly from Theorem 6. 19 and Theorem 5. 18.   #

Theorem 6. 28 Let s^∞ be periodic with period 2^n. If $L(s^\infty) < 2^n$, then $WP_{2k+1}(s^\infty) = 2^n$.

Proof: It follows directly from Theorem 6. 19 and Theorem 5. 21.   #

Theorem 6. 29 Let s^∞ be periodic with period 2^n. If $L(s^\infty) = 2^n$ and $0 < k < 2^{n-1}$, then $WP_{2k}(s^\infty) = 2^n$.

Proof: It follows directly from Theorem 6. 19 and Theorem 5. 22.   #

7 Summary and Open Problems

7.1 Summary and Open Problems of the Key Streams and Key Stream Generators

In this research report, by introducing the BAA attacks on several classes of stream ciphers, we have seen that the linear-complexity stability of key streams is of critical importance to stream ciphers. Actually, the BAA attacks are a kinds of correlation attack which make use of spectral techniques together with other algebraic methods. As shown in Section 3.3, the attacks can be fairly efficient in some cases. Being a kind of correlation analysis, the BAA attack is not limited to stream ciphers. It may be useful in block ciphers, especially in public key cryptosystems.

Correlation-immune functions have been introduced to stream ciphers as combining and filtering functions. They do protect stream ciphers from the divide and conquer attack. But in many cases they cannot protect stream ciphers from the BAA attacks. In fact, the BAA attack is more broad and more powerful, as the BAA attack make use of the correlation between a Boolean function and all linear functions. By employing the Xiao-Massey theorem and the energy conservation law, we have concluded that a lot of correlation-immune functions may not be the ideal combining and filtering functions for stream ciphers. At least, functions with high correlation-immune order are not preferable. In order to evaluate them further, we have to develop tight bounds on $PV(f)$ or $VS(f)$, which are measure indexes on the stability of Boolean functions. The second spectra of Boolean functions are not only governed by the energy conservation law, but also constrained by the following $2^n - 1$ equations:

$$\sum_{w \in GF(2)^n} S_{(f)}(w) S_{(f)}(w + v) = 0, \qquad v \neq 0$$

Thus it is reasonable to expect that $PV(f)$ and $VS(f)$ of Boolean functions with correlation—immunity would be larger because of the Xiao-Massey theorem and the above constraining equations. Anyway, the following problem is still an open problem.

Open Problem (I) How to develop tight bounds on $PV(f)$ and $VS(f)$ for correlation-immune Boolean functions.

In order to protect a stream cipher from the BAA attack, functions with flat spectra were introduced as combining or filtering functions for stream ciphers. Bent functions have not only the merit of having equal distances with all linear functions, but also of having a good autocorrelation property, i. e. , $C_f(w) = 1$ if $w = 0$, and $C_f(w) = 0$ if $w \neq 0$, where

$$C_f(w) = 2^{-n} \sum_{x \in GF(2)^n} (-1)^{f(x)+f(x \oplus w)}, \qquad w \in GF(2)^n$$

Nevertheless, they have the demerit that their nonlinear order is relatively small. Thus, the investigation of Boolean functions which are " nearly bent" is necessary. By "nearly bent" we mean that the absolute values of a Boolean function are approximately the same.

As shown in Section 3. 3, that the BAA attack is sometimes fairly efficient is due to the fact that the linear complexity of key streams is not stable. Therefore, we have introduced weight complexity and sphere complexity as measure indexes for the stability of linear complexity of sequences. It is obvious that we hope both the linear complexity $L(s^{\infty})$ and the sphere complexity $SC_u(s^{\infty})$ will be large while the sequence s^{∞} is employed as the key stream. But to achieve large values for both $L(s^{\infty})$ and $SC_u(s^{\infty})$ may be impossible. We believe there should be a trade-off between the linear complexity $L(s^{\infty})$ and the sphere complexity $SC_u(s^{\infty})$, and that the following conjecture is true.

Conjecture: For any binary sequence with period N , the following inequality holds

$$SC_u(s^\infty) + L(s^\infty) \leqslant N(1 + \frac{1}{u}) - 1$$

Example 7. 1 Let $s^\infty = (0\ 0 \cdots 01)^\infty$ with period N. Then it is apparent that $SC_1(s^\infty) = 0$ and $L(s^\infty) = N$. So we have $SC_1(s^\infty) + L(s^\infty) = N < 2N + 1$

Example 7. 2 Let $N = 2^n$, $u = 1$, and

$$S^N(x) = (1 + x)t(x)$$

where $t(1) \neq 0$ and $deg(t(x)) \leqslant 2^n - 2$. Then it follows from Theorem 5. 7 that $SC_1(s^\infty) = 2^n$. Thus

$$L(s^\infty) + SC_1(s^\infty) = 2^{n+1} - 1 = N(1 + \frac{1}{u}) - 1$$

This example shows that the above upper bound is achievable. However, although the foregoing conclusion is a conjecture, the following problem is still open.

Open Problem (**II**) Whether there is a trade-off between the linear complexity $L(s^\infty)$ and the sphere complexity $SC_u(s^\infty)$.

It follows from the results of Section 4. 4 that a stream cipher is not secure if its combining or filtering function is not stable. Integer addition has been employed in both the knapsack public cryptosystem and the knapsack stream ciphers. It is known that integer addition is a crytographically useful function, since its $GF(2)$ interpretations have high nonlinear orders. From the results of Section 4. 5 we can see that integer addition is cryptographically useful only when functioning as a whole since most of its $GF(2)$ -interpretation functions are either not stable or of lower nonlinear order.

To secure a stream cipher, it is necessary to make sure that its key stream has good linear-complexity stability. Thus the development of bounds on the weight complexity and sphere complexity and the determination of them for sequences are urgent. Thus we have developed some bounds on the weight complexity of binary sequences with period 2^n. The results show that some of them do have bad linear-complexity stability. ML-sequences are often used as basic components for constructing key streams, so it is essential to have a deep understanding of their linear-complexity stability. For this purpose we have developed lower bounds on the weight complexity of them. Those bounds tell us that their linear-complexity stability is really bad. They also tell us that to make a key stream constructed from ML-sequences with large linear complexity is fairly simple, but its linear—complexity stability may be of great defect. Hence one has to be careful when using them to construct key streams for stream ciphers. Clock-controlled sequences are another type of basic sequence used for constructing key streams, therefore it is imperative to develop bounds on the weight complexity and sphere complexity of them. Thus we have built some bounds on them. The results show that their linear complexity is relatively "good". Anyway, it is similar to that of ML-sequences. Fortunately, their linear complexity is inherently large. We can conclude that the kind of stream cipher presented in Fig. 5. 5 is hopeful and worth investigating. Although the statistical properties of clock-controlled sequences are not good, we can choose the filtering function mainly to control the statistical properties of their output sequences together with consideration of $PV(f)$ and the linear complexity of its output sequences. Nevertheless, this still remains to be investigated further.

Although we have developed many bounds on the weight complexity of many kinds of sequences, these bounds may not be tight. One may hope to determine the weight complexity and sphere complexity of some sequences. Generally speaking, this is rather difficult. We conjecture that it is a NP-complete problem. However, the following problem is still an open one.

Open Problem (III)   Whether the problem of determining the weight complexity $WC_u( )$ and sphere complexity $SC_u(\cdot)$ for a general sequence with period N is NP-complete.

If the problem of determining weight complexity or sphere complexity is an NP-complete one, this does not means that the problem cannot be investigated further. The determination of the weight or sphere complexity for some special sequences is possible, and at least developing tight bounds for the weight complexity or sphere complexity of some sequences is feasible. The problem of decoding linear codes is a good example to illustrate this.

Complexity theory classifies a problem according to the minimum time and space needed to solve the hardest instances of the problem on a Turing Machine(or some other abstract model of computation). A Turing Machine(TM) is a finite state machine with an infinite read-write tape (see [Trau 88] [Gare 79]). A TM is a "realistic" model of computation in that problems that are polynomial solvable on a TM are also polynomial solvable on real systems and vice versa.

Problems are called tractable if they are solvable in polynomial time and they can usually be solved for reasonable size inputs. Intractable or hard problems are those that cannot be systematically solved in polynomial time, because as the size of the input increases, their solution becomes infeasible on even the fastest computers. The class P consists of all tractable problems, that is, all problems solvable in polynomial time on a deterministic computer. The class NP (nondeterministic polynomial) consists of all problems solvable in polynomial time on a nondeterministic TM. This means if the machine guesses the solution, it can check its correctness in polynomial time. Of course, this does not really "solve" the problem, because there is no guarentee the machine will guess the right answer. An example of such a problem is the " Knapsack Problem" : given a set of n integers $A = \{a_1, a_2, \ldots, a_n\}$ and an integer S, to determine whether there exists a subset of A that sums to S. It is easy to see that the problem is in NP because for any given subset, it is easy to check whether it sums to S. Finding a subset that sums to S is much harder, as there are 2^n possible subsets; trying all of them has time complexity $O(2^n)$.

The class NP includes class P because any problem polynomial solvable on a deterministic TM is polynomial solvable on a nondeterministic one. If all NP problems are polynomial solvable on a deterministic TM, we would have $P = NP$. Although many problems in NP

seem much "harder" than the problems in P, no one has yet proved that $P \neq NP$. The class NP-complete is the set of equivalent problems such that if any one of the problems is in P then $P = NP$. Thus, the NP-complete problems are the "hardest" problems in NP. More and more problems have been shown to be in NP-complete. The knapsack problem belongs to the class of NP-complete problems [Gare 79].

As another approach to the stability of linear complexity, we have introduced two other measure indexes on the linear-complexity stability, that is, fixed-complexity distance (FCD) and variable-complexity distance (VCD). By employing Blahut's Theorem, the relationships between weight complexity and fixed-complexity distance as well as sphere complexity and variable-complexity distance have been established. It is a pity that we have not given many bounds on the VCD and FCD of periodic sequences. One may find sequences with good linear-complexity stability if one can build tight bounds on them. Since maximum-length sequences and M-sequences as well as clock-controlled sequences are elementary sequences for building key streams, it is necessary to develop bounds on VCD and FCD for these sequences. Thus, we propose the following open problem.

Open Problem (**IV**)   How to determine $FCD_k(\cdot)$ and $VCD_k(\cdot)$ or to build tight bounds on them for m-sequences, M-sequences and clock-controlled sequences as well as other kinds of periodic sequences.

As shown in Chapter 6, the linear complexity of a sequence is not stable if its period is not stable. This is due to the fact that the linear-complexity stability of a sequence is strongly related to its period stability. Weight period and sphere period were introduced to measure the period stability of sequences. Unfortunately, we have not developed many bounds on weight period $WP_k(\cdot)$ and sphere period $SP_k(\cdot)$, because the development of bounds on $WP_k(\cdot)$ and $SP_k(\cdot)$ is much more difficult than that on $WC_k(\cdot)$ and $SC_k(\cdot)$. Since ML-sequences, M-sequences and clock-controlled sequences are the basic sequences for building key streams, it is important to develop bounds on $WP_k(\cdot)$ and $SP_k(\cdot)$.

Thus, we propose the following open problem.

Open Problem (V) How to determine tight bounds on $WP_k(\cdot)$ and $SP_k(\cdot)$ for ML-sequences, M-sequences and clock-controlled sequences as well as other kinds of sequences which are useful in stream ciphers.

Let s^{∞} be a sequence with period N and satisfy the following nonlinear equation

$$s_n = f(s_{n-1}, s_{n-2}, \ldots, s_{n-L}), \qquad n > L$$

where $f(x_1, x_2, \ldots, x_L)$ is a nonlinear function. If we can find a linear function $L(x_1, \ldots, x_L)$ and a periodic sequence with period K such that

$$t_n = L(t_{n-1}, \ldots, t_{n-L})$$

and

$$p = \frac{weight(s_0 - t_0, s_1 - t_1, \ldots, s_{M-1} - t_{M-1})}{M}$$

is very small, where $M = lcm(N, K)$, then the linear-complexity stability of s^{∞} is very bad. One may expect $L(x_1, x_2, \ldots, x_L)$ to be the best affine approximation of $f(x_1, \ldots, x_L)$. Actually, this is not generally the case. For example, let

$$f(x_1, x_2, \ldots, x_6) = x_1 + x_2 + x_3 x_4 x_5 + x_6$$

and $s_0 s_1 s_2 s_3 s_4 s_5 = 100101$, then s^{∞} has period 36 and

$$s^{36} = 100101011100100110000101101010001110$$

Or let $L(x) = x_1 + x_2 + x_6$ and $t_0 t_1 t_2 t_3 t_4 t_5 = 100101$ as well as t^{∞} satisfy $t_n = L(t_{n-1}, \ldots, t_{n-6})$. Then t^{∞} has period 31 and

$$t^{31} = 1001010111000100000110110011110$$

It is easy to see that $L(t^{\infty}) = 6$. Noticing that the Hamming weight of s^{36} is 17, which is odd, it follows from Theorem 5.3 that $L(t^{\infty}) = 36$. Let $M = lcm(\text{period } s^{\infty}, \text{period } t^{\infty}) = 36 \times 31 = 1116$. Since the probability of agreement between $f(x)$ and $L(x)$ is $7/8$, one may expect that of agreement between s^{∞} and t^{∞} to be large. Actually, $p = 559/1116 \approx 0.5009$. This is due to the fact that there are error propagations when using $L(x)$ as a substitute for $f(x)$ to iterate s^{∞}. This problem seems very difficult. Because of its importance, we propose the following problem.

Open Problem (VI) With the assumptions as above, how to find the linear function $L(x_1, \ldots, x_L)$ and the sequence t^{∞} such that p is minimal.

Open Problem (VII) It is known that for a binary de Bruijn sequence of period 2^n, for $n \geqslant 3$, its linear complexity lies between $2^{n-1} + n$ and $2^n - 1$. Since de Bruijn sequences have maximum period and large linear complexity, we propose the following problem for cryptographical purpose:

(a) find the stability of Boolean functions of the feedback shift registers which producd the de Bruijn sequences.

(b) determine the linear-complexity stability of de Bruijn sequences.

It can be seen that the determination of the best affine approximation of functions from Z_p^n to Z_p is useful in estimating the linear-complexity stability of some decimated sequences. Thus, it is cryptographically important to solve the following problem.

Open Problem (VIII) How to determine the best affine approximation of functions from Z_p^n to Z_p.

Since maximum-length sequences of period 2^n have linear complexity greater than or equal to $2^{n-1} + n$ for $n \geqslant 3$, they are cryptographically attactive. Because (N, K)-maximum-length sequences resemble maximum-length sequences, they may be cryptographically useful. Thus, it is worthwhile to investigate the following problem.

Open Problem (IX) Specify the (N, K)-maximum-length sequences [for a definition see Section 6. 2]

7. 2 On the Stability of Source Coding for Binary Additive Stream Ciphers

In the foregoing chapters, we have disscussed the stability of key streams and key stream generators. In this section, we shall illustrate the cryptographic importance and the stability of source coding for some stream ciphers. Let us first consider the synchronous binary additive stream cipher in Fig. 7. 1, where $c_t = m_t \oplus z_t$, and z^∞ is the binary key stream. The purpose of adding the key stream to the plaintext sequence is to conceal the plaintext bits. This means that the function of the key stream is to protect the plaintext. Nevertheless, from the expression $c_t = m_t \oplus z_t$, we see that m_t and z_t are equally important from the viewpoint of their contributions to c_t. Thus, while designing a stream cipher such as in Fig. 7. 1, we should not only consider the problem of how to employ the key stream to protect the plaintext, but also ponder the problem of source coding for the purpose of protecting the key stream by coded plaintext bits. In other words, in order to be protected better by the key stream, the coded plaintext bits should have the function of protecting the key stream. Taking an extreme example, suppose that a discrete memoryless source U has a k letter alphabet $a_1, \ldots, a_k$ with probabilities $p(a_1), \ldots, p(a_k)$. Each source letter is to be represented by a finite binary sequence as in Table 7. 1.

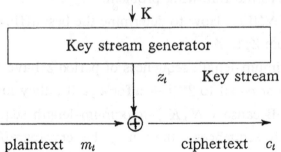

Fig. 7. 1　The binary additive stream cipher

The variable-length code in the table is not a prefix condition one, but is uniquly decodable. If the code is used for the source code of the source of the binary additive stream cipher of Fig. 7. 1, then $P(z_t = c_t \oplus 1)$ must be very large, since $P(z_t = c_t \oplus 1) = P(m_t = 1)$ which is also large. Consequently, most components of the key stream must be exposed in the ciphertext sequence. Thus, for designing the secure binary stream cipher of Fig. 7. 1, it is profitable to develop a source coding method by which the probability $P(m_t = 1)$ can be made as closely $\frac{1}{2}$ as possible.

Table 7. 1

Source Letter	Probability	Code
a_1	$P(a_1)$	1
a_2	$P(a_2)$	01
$\vdots$	$\vdots$	$\vdots$
a_k	$P(a_k)$	$0\cdots01$
		$\underbrace{}$
		$k-1$

We now derive the above cryptographic requirement from another point of view. For the above binary additive stream cipher, we regard m_t, z_t, c_t as three binary random variables. Because of the structure of the stream cipher, we make the following assumptions:

1) m_t and z_t are statistically independent;
2) $P(z_t = 1) = \alpha,\quad 0 \leqslant \alpha \leqslant 1$;
3) $P(m_t = 1) = \beta,\quad 0 \leqslant \beta \leqslant 1$.

Therefore, we get

$$P(c_t = 1) = P(m_t = 0)P(z_t = 1) + P(m_t = 1)P(z_t = 0)$$
$$= (1 - \beta)\alpha + \beta(1 - \alpha)$$
$$= \alpha + \beta - 2\alpha\beta \qquad (1)$$

Denoting $\gamma = P(c_t = 1)$, we obtain

$$\gamma = \alpha + \beta - 2\alpha\beta \qquad (2)$$

whence,

$$\gamma - \frac{1}{2} = -2(\alpha - \frac{1}{2})(\beta - \frac{1}{2}) \qquad (3)$$

$$|\gamma - \frac{1}{2}| = 2|\alpha - \frac{1}{2}||\beta - \frac{1}{2}| \qquad (4)$$

Thus, $\gamma = \frac{1}{2}$ if and only if $\alpha = \frac{1}{2}$ or $\beta = \frac{1}{2}$. Suppose a cryptanalyst only get the information that $\beta > \frac{1}{2}$, then by formula (3), we get

$$\gamma - \frac{1}{2} = 2(\frac{1}{2} - \alpha)a \qquad (5)$$

where $a = \beta - \frac{1}{2} > 0$. Consequently, if the number of 1s in a segment of ciphertext sequence is much larger (smaller) than that of 0s, then γ must be much larger (smaller) than $\frac{1}{2}$. Therefore, by formula (5) we see α must be much smaller (larger) than $\frac{1}{2}$. Thus, the cryptanalyst get the information that there are much fewer (more) 1s in the corresponding segment of the plaintext sequence. This also shows that to develop a source coding method such that $P(m_t = 1)$ is as closely $\frac{1}{2}$ as possible is cryptographically profitable. One can also arrive at the above conclusion by analyzing the mutual information between c_i and m_t.

From the above analysis, the number of 1s and that of 0s in a plaintext sequence should be roughly equal. We now introduce an index for the measurement of the extent of balance between 0s and 1s. Assume that a discrete memoryless source U has k letter alphabet $a_1, \ldots, a_k$ with probabilities $P(a_1), \ldots, P(a_k)$, and a_i is coded as a binary codeword c_i with length n_i for $1 \leqslant i \leqslant k$. Let m_i denote the number of 1s in c_i, $1 \leqslant i \leqslant k$. We take the following index as a measurement for the stability of the source code C:

$$P_C = \sum_{i=1}^{k} p(a_i) \frac{m_i}{n_i}$$

P_C can be regarded as a rough measurement for the probability of appearance of 1 in a position of a segment of coded source alphabets. We call a source code C unconditionally stable if $P_C = \frac{1}{2}$. For every source U as above, there do exist some unconditionally stable codes. For example, choose an even integer n such that $\binom{n}{n/2} > k$, then assign vectors of $GF(2)^n$ with Hamming weight $n/2$ to each alphabet of U. Apparently, the above source code is unconditionally stable. However, the codeword length of the above fixed-length source code is much larger than that of its optimum code. Generally speaking, it seems that there should exist a trade-off between P_C and the average codeword length $\bar{n}$.

There are two kinds of source codes: the fixed-length codes and variable-length codes. We call a fixed-length source code C of length n conditionally stable if P_C satisfies

$$|P_C - \frac{1}{2}| = \min\{ |P_D - \frac{1}{2}| : D \text{ is a source code for } U \text{ with}$$

$$\text{codeword length } n \}$$

Similarly, a variable-length code with average length $\bar{n}$ is said to be conditionally stable if P_C satisfies

$$|P_C - \frac{1}{2}| = \min\{ |P_D - \frac{1}{2}| : D \text{ is a variable-length code for } U$$

$$\text{with average codeword length } \bar{n} \}$$

Since it is not known how to construct conditionally stable source codes, we propose the following problem:

Open Problem (X) A discrete memoryless source U has k letter alphabet $a_1, \ldots, a_k$ with probabilities $P(a_1), \ldots, P(a_k)$.

(a) Given length n, how to construct a conditionally stable source code C with fixed codeword length n for source U.

(b) Given a positive integer $\bar{n}$, how to construct a conditionally stable source code C with average codeword length $\bar{n}$ for source U, where

$$\bar{n} = \sum_{i=1}^{k} P(a_i) n_i$$

and n_i is the length of the codeword for the alphabet a_i.

We have already seen that employing conditionally stable codes for the plaintext sources of binary additive stream ciphers is cryptographically beneficial, but our analysis may be superficial. However, we be-

lieve that the study of source coding techniques for ciphers is worthwhile, and should be investigated further.

Appendix A

Massey's Conjectured Algorithm for the Linear Feedback Shift Register Synthesis of Multi-sequences and Its Applications

A. 1　Massey's Conjectured Algorithm

It is well known that the SLFSR (shortest linear feedback shift register) synthesis of a single sequence is of great importance in practice [Mass 69], especially in cryptology and coding theory. Since there is an efficient algorithm for the SLFSR synthesis of a single sequence, i. e. , the well known Berlekamp-Massey algorithm, the linear complexity of a key stream becomes an important measure index on the strength of a stream cipher. The problem of synthesizing multi-sequences with LFSR has been much studied by many scholars in information and control theory for many years. Let $B_i^N = a_{i1} \cdots a_{iN}$, $i = 1, \cdots, M$, be M sequences of length N in a field F , and $s_i = (a_{i1}, a_{i2}, \cdots, a_{im})^t$ as well as $S = (B_1, B_2, \cdots, B_M)^t$. The determination of a polynomial $f(x) \in F[x]$ and the smallest positive integer L , where

$$f(x) = c_0 + c_1 x + \cdots + c_L x^L, \qquad c_0 = 1, \ c_i \in F$$
$$i = 0, 1, \cdots, L$$

such that

$$s_{i+1} + c_1 s_i + \cdots + c_L s_{i-L+1} = 0, \qquad L \leqslant i \leqslant N - 1$$

is called the SLFSR synthesis of the sequences B_1^N , $\cdots$, B_M^N . Since a linear feedback shift register has two parameters, the feedback polynomial $f(x)$ and the length L, we shall use $(f(x), L)$ to denote a LFSR. Let $s^N = s_1 s_2 \cdots s_N$ be a sequence of length N; we shall use s^i to express the finite sequence $s_1 s_2 \cdots s_i$.

In 1972, J. L. Massey gave a conjectured algorithm for the SLF-SR synthesis of multi-sequences [Mass 85]. G. Fen and K. K. Tzeng gave another one in 1985 [Fen 85]. Massey's conjectured algorithm was proven not only to be true, but also to be a universal one suitable for the minimal realization of any linear system by Ding in 1987 [Ding 87]. Massey's algorithm is depicted in Fig. A. 1, and can be stated as in the following theorem.

Massey's Conjectured Theorem: Let (f_i, l_i) be the SLFSR which generates s_i , and $d_i = f_i(s^{i+1})$ be the ith discrepancy, $i = 0, 1, \cdots, n$. Then

(i) If $d_n = 0$, then $l_{n+1} = l_n$ and $f_{n+1} = f_n$.

(ii) If $d_n \neq 0$, and is a linear combination of $d_0, d_1, \cdots, d_{n-1}$, let $d_{k_1}, d_{k_2}, \cdots, d_{k_r}$ be a basis of $\{d_i: 0 \leqslant i \leqslant n - 1\}$ such that if

$$d_i = u_1 d_{k_{j(1)}} + \cdots + u_p d_{k_{j(p)}},$$

$u_j \neq 0$ for $j = 1, 2, \cdots, p$ and $\{j(1), \cdots, j(p)\} \in \{1, 2, \cdots, r\}$, then

$$n - i + l_i \geqslant max\{u - k_{j(i)} + l_{k_{j(i)}}: 1 \leqslant i \leqslant p\}$$

and $(k_1, \cdots, k_r)$ is maximal in alphabetic order. Let

$$d_n = - \sum_{i=1}^{r} u_i d_{k_i}$$

$$I = \{i: u_i \neq 0, 1 \leqslant i \leqslant r\}$$

then

$$l_{n+1} = max\{l_n, max\{n - k_i + l_{k_i}: i \in I\}\}$$

$$f_{n+1} = f_n + \sum_{i=1}^{r} u_i x^{n-k_i} f_{k_i}$$

(iii) If d_n is not a linear combination of $d_0, d_1, \cdots, d_{n-1}$, then $l_{n+1} = n + 1$ and f_{n+1} can be any polynomial of degree $n + 1$ in $F[x]$.

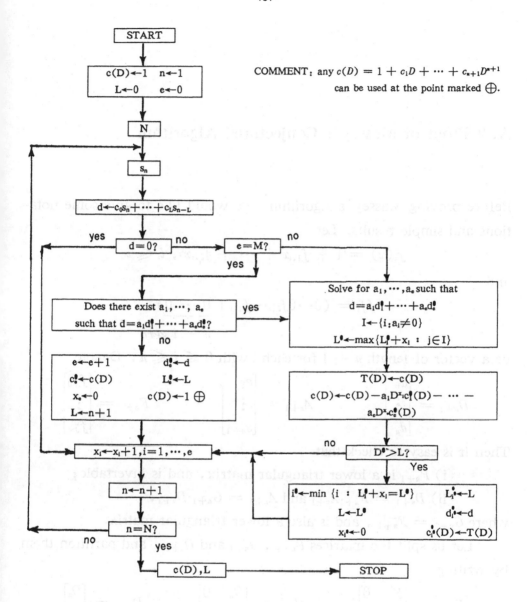

COMMENT: any $c(D) = 1 + c_1 D + \cdots + c_{n+1} D^{n+1}$ can be used at the point marked $\oplus$.

Fig. A. 1 Massey's algorithm for the LFSR synthesis of multi-sequences

A. 2 Proof of Massey's Conjectured Algorithm

Before proving Massey's algorithm, we would like to give some notations and simple results. Let

$$f_i(x) = 1 + f_{1,i}x + \cdots + f_{l_i,i}x^{l_i}, \quad l_i \leqslant i$$

and

$$ff_i = (0\cdots 0 f_{l_i,i}\cdots f_{1,i}\, 1\, \underbrace{0\, \cdots\, 0})_{n - i\ zero}$$

be a vector of length $n + 1$ for each i with $0 \leqslant i \leqslant n$. Denote

$$D_{n+1} = \begin{bmatrix} d_0 \\ \vdots \\ d_n \end{bmatrix} \qquad A_{n+1} = \begin{bmatrix} s_1 \\ \vdots \\ s_{n+1} \end{bmatrix} \qquad F_{n+1} = \begin{bmatrix} ff_0 \\ \vdots \\ ff_n \end{bmatrix}$$

Then it is easy to check that

(i) F_{n+1} is a lower triangular matrix, and is invertable;

(ii) $D_{n+1} = F_{n+1}A_{n+1}$ and $A_{n+1} = G_{n+1}\cdot D_{n+1}$,

where $G_{n+1} = F_{n+1}^{-1}$, and is also a lower triangular matrix.

Let us split the matrices F_{n+1}, G_{n+1} and D_{n+1}, and partition them by writing

$$F_{n+1} = \begin{bmatrix} F_n & 0 \\ r_n & 1 \end{bmatrix} \qquad G_{n+1} = \begin{bmatrix} G_n & 0 \\ g_n & 1 \end{bmatrix} \qquad D_{n+1} = \begin{bmatrix} D_n \\ d_n \end{bmatrix}$$

where F_n and G_n are two $n \times n$ matrices, and

$$D_n = \begin{bmatrix} d_0 \\ \vdots \\ d_{n-1} \end{bmatrix}$$

Define matrix $U_{(n-L+1)\times(n+1)}$ as

$$\begin{bmatrix} 0 & 0 & \cdots\cdots & 0 & u_l & \cdots\cdots & u_1 & 1 \\ 0 & 0 & \cdots\cdots & u_L & u_{L-1} & \cdots\cdots & 1 & 0 \\ \vdots & \vdots & & \vdots & \vdots & & \vdots & \vdots \\ \vdots & \vdots & & \vdots & \vdots & & \vdots & \vdots \\ 0 & u_L & \cdots\cdots & u_1 & 1 & 0 \cdots & 0 & 0 \\ u^L & u^{L-1} & \cdots\cdots & 1 & 0 & 0 \cdots & 0 & 0 \end{bmatrix}$$

and partion it by writing

$$U_{(n-L+1)\times(n+1)} = \begin{bmatrix} B & 1 \\ U_{(n-L)\times n} & 0 \end{bmatrix}$$

where $B = (0 \cdots 0 \; u_L \cdots u_1)$ is a vector of length n , and $0 = (0 \cdots 0)^t$ is a vector of length $n - L$.

Theorem A. 1 Let $f(x) = 1 + u_1 x + \cdots + u_L x^L$ with $L < n + 1$, then $(f(x), L)$ generates s^{n+1} if and only if $U_{(n-L)\times n} G_n D_n = 0$ and $BG_n D_n + g_n D_n + d_n = 0$.

Proof: Since LFSR $(f(x), L)$ generates s^{n+1} and $L < n + 1$, by definition it is trivial to check that Theorem A. 1 is valid.   #

Theorem A. 2 If $(f(x), L)$ can generate s^{n+1} and $L < n + 1$, then there must exist a vector $u = (u_0, u_1, \cdots, u_{n-1})$ such that

$$f(x) = f_n(x) + \sum_{i=0}^{n-1} u_i x^{n-i} f_i(x)$$

and

$$d_n = - uD_n = - \sum_{i=0}^{n-1} u_i d_i$$

Proof: By Theorem A. 1, we get $BG_n D_n + g_n D_n + d_n = 0$. Since $F_{n+1} G_{n+1} = I_{n+1}$, where I_{n+1} is the $(n+1) \times (n+1)$ identity matrix, so $r_n = - g_n F_n$. Let $BG_n + g_n = u$, then $B = - g_n F_n + uF_n = r_n + uF_n$. This shows

$$f(x) = f_n(x) + \sum_{i=0}^{n-1} u_i x^{n-i} f_i(x)$$

On the other hand, we have

$$d_n = - (BG_n + g_n)D_n = - uD_n$$

$$= - \sum_{i=0}^{n-1} u_i d_i \qquad \#$$

Theorem A. 3 Let $(f_i(x), l_i)$ be the SLFSR which generates s^i and d_i be the i th discrepancy, $i = 0, \cdots, n$, then $l_{n+1} = n + 1$ if and only if d_n is not a linear combination of d_i, $i = 0, \cdots, n - 1$.

Proof: If d_n is not a linear combination of d_i, $i = 0, 1, \cdots, n - 1$, then it follows directly from Theorem A. 2 that $l_{n+1} = n + 1$. Conversely, if $l_{n+1} = n + 1$, suppose d_n is a linear combination of d_0, d_1, $\cdots$, d_{n-1}, say,

$$d_n = - \sum_{i=0}^{n-1} u_i d_i$$

and set

$$f(x) = f_n(x) + \sum_{i=0}^{n-1} u_i x^{n-i} f_i(x)$$

Noticing that $l_i \leqslant i$, we get $deg(f(x)) \leqslant n - 1$. Therefore, it is easy to check that $(f(x), n)$ is a LFSR that can generate s^{n+1}. Thus $l_{n+1} \leqslant n$. This is contrary to $l_{n+1} = n + 1$. Hence, d_n is not a linear combination of d_0, $d_1, \cdots$, d_{n-1}. $\qquad \#$

Let $f(x)$ be a polynomial of degree less than or equal to n, then there must exist at least one positive integer L with $L \leqslant n$ such that $(f(x), L)$ generates s^n. For instance, $(f(x), n)$ is such a LFSR. Thus, there must exist the smallest L such that $(f(x), L)$ generates s^n with $L \geqslant deg(f)$. Here and hereafter, we shall use L' to denote the smallest L, and refer to it as the associated integer of $f(x)$ with respect to s^n.

Lemma A. 4 Let (f_m, l_m) and (f_{k_1}, l_{k_1}) be two shortest LFSRs that generate s^m and s^{k_1} respectively, and l'_m and l'_{k_1} be the associated integers of $f_m(x)$ and $f_{k_1}(x)$ respectively with respect to s^m and s^{k_1}. Assume $k_1 < m$ and set

$$g(x) = f_m(x) + u_1 x^{m-k_1} f_{k_1}(x), \quad u_1 \neq 0$$

Then that $(g(x), L)$ generates s^m implies

$$L \geqslant max\{l'_m, m - k_1 + l'_{k_1}\}$$

$$= max\{l_m, m - k_1 + l_{k_1}\}$$

Proof: It follows from the definitions of (f_m, l_m) and (f_m, l_m') that $l_m' = l_m$. Similarly, we get $l_{k_1} = l_{k_1}'$. Since $L \geqslant l_m$, so $L \geqslant l_m'$. Suppose $l_m' \leqslant L \leqslant m - k_1 + l_{k_1}'$. Let j be the last j such that $f_{k_1, j} \neq 0$.

Case (1) If $j + m - k_1 \leqslant l_m'$, because $L \geqslant l_m'$, so we have $L - m + k_1 \geqslant l_m' - m + k_1 \geqslant j$. Put $LL = L - m + k_1$ and $h(x) = 1 + h_1 x + \cdots + h_j x^j$, where $h_i = f_{k_1, i}$, $i = 1, \cdots, j$. Then we get

$$g(x) = f_m(x) + u_1 x^{m - k_1} h(x), \quad j \leqslant LL < l_{k_1}'$$

Since (g, L) generates s^m and $L \geqslant l_m$, so we have $g(s^m) = \cdots = g(s^{L+1}) = 0$. Similarly, it follows that $f_m(s^m) = \cdots = f_m(s^{L+1}) = 0$. Thus, we get

$$h(s^{k_1}) = \cdots = h(s^{LL+1}) = 0$$

This means that $(h(x), LL) = (g(x), LL)$ can generate s^{k_1} with $LL < l_{k_1}'$. This is contrary to the minimality of l_{k_1}'. Hence $L \geqslant m - k_1 + l_{k_1}' = max\{l_m', m - k_1 + l_{k_1}'\} = max\{l_m, m - k + l_{k_1}\}$.

Case (2) If $j + n - k_1 > l_m'$, regard $g(x)$ and $f_m(x) + x^{m - k_1} f_{k_1}(x)$ as polynomials of degree m, $m \leqslant n$. Then the number of degenerating terms of $f_m(x) + x^{m - k_1} f_{k_1}(x)$ is $m - (m - k_1 + j)$, so it follows that $m - L \leqslant m - (m - k_1 + j)$. Put $LL = L - m + k_1$, then $j \leqslant LL < l_{k_1}'$. For the same reason as in case (1) we know that $(h(x), LL)$ generates s^{k_1} with $LL < l_{k_1}'$. This is also contrary to the minimality of l_{k_1}'. Thus, we get

$$L \geqslant max\{l_m', m - k_1 + l_{k_1}'\}$$
$$= max\{l_m, m - k_1 + l_{k_1}\}$$

By taking Cases (1) and (2) together, we see that Lemma A. 4 is valid. $\#$

Theorem A. 5 Let $(f_i(x), l_i)$ be a shortest LFSR that generates s^i, $i = 1, \cdots, n$, and l_i' be the associated integer of $f_i(x)$ with respect to s^i. Set

$$g(x) = f_n(x) + \sum_{i=1}^{t} u_i x^{n - k_i} f_{k_i}(x), \quad u_i \neq 0, \quad i = 1, \cdots, t$$

If $(g(x), L)$ generates s^n, then

$$L \geqslant max\{l_n, n - k_1 + l_{k_1}, \cdots, n - k_t + l_{k_t}\}$$

Proof: By induction on t. It follows from Lemma A. 4 that Theorem A. 5 is true for $t = 1$. Suppose it holds for $t - 1$ and consider case t. If there are two (or more) integers $n - k_i + l_{k_i}$ and $n - k_j + l_{k_j}$ such that $i \neq j$, $i, j \in \{1, 2, \cdots, t\}$ and

$$n - k_i + l_{k_i} = n - k_j + l_{k_j}$$
$$= max\{n - k_1 + l_{k_1}, \cdots, n - k_t + l_{k_t}\}$$

then, without loss of generality, assume $k_i < k_j$. Set

$$h(x) = f_{k_j}(x) + u_i u_j^{-1} x^{k_j - k_i} f_{k_i}(x)$$

Noticing that $l_{k_j} = k_j - k_i + l_{k_i}$, we can check that $(h(x), l_{k_j})$ is another shortest LFSR which generates the sequence s^{k_j} by writing $g(x)$ as

$$g(x) = f_n(x) + u_j x^{n-k_j} h(x) + \sum_{v \neq i, j} u_v x^{n-k_v} \cdot f_{k_v}(x)$$

Therefore, it follows from the induction hypothesis that

$$L \geqslant max\{l_n, \{n - k_v + l_{k_v}, v \neq i\}\}$$
$$= max\{l_n, n - k_j + l_{k_j}\}$$
$$= max\{l_n, n - k_1 + l_{k_1}, \cdots, n - k_t + l_{k_t}\}.$$

If there is only one integer $n - k_s + l_{k_s}$ such that $1 \leqslant s \leqslant t$ and

$$n - k_s + l_{k_s} = max\{n - k_1 + l_{k_1}, \cdots, n - k_t + l_{k_t}\}.$$

then for the case that $l_n \geqslant n - k_s + l_{k_s}$, it follows from $l_{n+1} \geqslant l_n$ that

$$L \geqslant max\{l_n, n - k_1 + l_{k_1}, \cdots, n - k_t + l_{k_t}\} = l_n$$

and for the case that $l_n < n - k_s + l_{k_s}$, noticing that

$$n - k_i + l_{k_i} < n - k_s + l_{k_s} \text{ for } i \neq s$$

and setting $p = n - k_s + l_{k_s}$, we get $p > l_n$ and

$$p > n - k_i + l_{k_i} \text{ for } i \neq s$$

Suppose $L < n - k_s + l_{k_s}$, then $p > L$. So it follows that $g(s^p) = 0$ and $f_{k_i}(s^{p-n+k_i}) = 0$ for $i \neq s$. Therefore, setting $q = l_{k_s}$ we get

$$u_s f_{k_s}(s^q) = g(s^p) - \sum_{i \neq s} u_i f_{k_i}(s^{p-n+k_i})$$
$$= 0$$

Hence, $(f_{k_s}(x), l_{k_s} - 1)$ can generate s^{k_s}. This is contrary to the minimality of l_{k_s}. Thus it must be the case that

$$L \geqslant n - k_s + l_{k_s}$$
$$= max\{l_n, n - k_1 + l_{k_1}, \cdots, n - k_t + l_{k_t}\}$$

Based on the induction hypothesis, we have so far proved that Theorem A. 5 is also true for t. By induction on t, we have completed the proof of Theorem A. 5.   #

Theorem A. 6 Let (f_i, l_i) be a shortest LFSR which generates s^i, and $d_i = f_i(s^{i+1})$, $i = 0, 1, \cdots, n - 1, n$. If $d_n \neq 0$ and can be expressed as a linear combination of d_i, $i = 0, 1, \cdots, n - 1$, say

$$d_n = - uD_{n-1} = - \sum u_i d_i$$

Set $I_u = \{0 \leqslant i \leqslant n - 1; u_i \neq 0\}$, then

$$l_{n+1} = \min_{d_n = -uD_n} max\{l_n, n - i + l_i; i \in I_u\} \qquad (1)$$

$$f_{n+1} = f_n + \sum_{i=0}^{n-1} u'_i x^{n-i} f_i$$

where $u' = (u'_0, \cdots, u'_{n-1})$ is a vector which makes the right side of (1) take its minimal value.

Proof: Let L denote the right side of (1). It is obvious that $l_{n+1} \leqslant L$. Since d_n can be expressed as a linear combination of $d_0, d_1, \cdots, d_{n-1}$, it follows from Theorem A. 3 that $l_{n+1} < n + 1$. Hence it follows from Theorem A. 2 and A. 5 that $l_{n+1} \geqslant L$. This proves Theorem A. 6.   #

Proof of Massey's Conjectured Theorem:

Part (i) of Massey's conjectured theorem is apparently true. Part (iii) follows from Theorem A. 3. We now prove part (ii). According to Theorem A. 6, we have

$$l_{n+1} = \min_{d_n = -uD_{n-1}} max\{l_n, n - i + l_i; i \in I_u\}$$

Let $B_n = \{d_{k_1}, \cdots, d_{k_r}\}$ be the basis chosen in Massey's algorithm in step n, and

$$d_n = - \sum_{i=1}^{s} u_i d_{k_{j(i)}}, \quad u_i \neq 0, i = 1, \cdots, s \qquad (2)$$

where $j(i) \in \{1, \cdots, r\}$, $i = 1, \cdots, s$. So we know that

$$l_{n+1}^M = max\{l_n, \{n - k_{j(i)} + l_{k_{j(i)}}; 1 \leqslant i \leqslant s\}\}$$

where l_{n+1}^M is the length calculated according to Massey's algorithm. It is obvious that $l_{n+1} \leqslant l_{n+1}^M$. Suppose that $l_{n+1} < l_{n+1}^M$, and

$$l_{n+1} = max\{l_n, n - h_i + l_{h_i} : 1 \leqslant i \leqslant t\}$$

$$d_n = \sum_{i=1}^{t} v_i d_{h_i}, \ v_i \neq 0, d_{h_i} \neq 0, \ i = 1, \cdots, t$$

If $l_{n+1}^M = l_n$, then $l_{n+1} \geqslant l_n = l_{n+1}^M$. Hence $l_{n+1} = l_{n+1}^M$. therefore, assume that $l_{n+1}^M = n - k_1 + l_{k_1}$. Since $l_{n+1} < l_{n+1}^M$, we get

$$n - h_i + l_{h_i} < n - k_1 + l_{k_1}, \ i = 1, \cdots, t$$

From the method for choosing the basis in Massey's algorithm we know that d_{h_i} must be a linear combination of $\{d_{k_2}, \cdots, d_{k_r}\}$. Therefore, $d_n = v_1 d_{h_1} + \cdots + v_t d_{h_t}$ must also be a linear combination of $d_{k_2}, \cdots, d_{k_r}$, say

$$d_n = w_1 d_{k_2} + \cdots + w_r d_{k_r} \tag{3}$$

By (2) and (3) we know that d_{k_1} must be a linear combination of $\{d_{k_2}, \cdots, d_{k_r}\}$. This is contrary to the linear independency of $\{d_{k_1}, \cdots, d_{k_r}\}$. Hence $l_{n+1} = l_{n+1}^M$.

Remark: From the proof of Massey's algorithm we know that the condition that $(k_1, \cdots, k_r)$ is maximal in alphabetic order in the algorithm can be removed from the algorithm. Thus, the speed of the algorithm can be improved.

In what follows in this section, we shall show that Massey's algorithm is a universal one. Let V be a vector space over the field F, and $s = s_1 s_2 \cdots s_n$ be a vector sequence of length N over field F. The problem of finding a pair $(f(x), l_n)$, where

$$f_n(x) = c_0 + c_1 x + \cdots + c_l x^{l_n},$$

$$c_0 = 1, c_i \in F, i = 1, \cdots, l_n.$$

such that $(f_n(x), l_n)$ generates s^n, i. e. ,

$$s_k + c_1 s_{k-1} + \cdots + c_l s_{k-l_n} = 0, \ n \geqslant k \geqslant l_{n+1},$$

and l_n is minimal, is referred to as the minimal realization for vector sequences. Notice that the above proofs do not require us to know what the s_i are, but only require that they belong to a vector space over F. So all the above results are true for vector sequences. This shows that

Massey's algorithm is a universal one, it is suitable for the minimal realization of any linear system. We now give some special cases of the universal algorithm:

1) If $V = F$, then it is the B-M algorithm;

2) If $V = F^m$, then it is Massey's algorithm for the LFSR synthesis of multi-sequences;

3) If $V = F_{n \times n}$, then it gives a minimal realization algorithm for matrix sequences.

4) If $F = GF(q)$, $V = GF(q^n)$, then it gives an algorithm for the lower field linear complexity $L_{GF(q)}(s)$ for the sequences over $GF(q^n)$.

A. 3　An Application of Massey's Algorithm to Cryptology

Let $M = (m_1, \cdots, m_n)$ be a block of message digits over $GF(q)$, and let A be a nonsingular $n \times n$ matrix over $GF(q)$. If a block cipher enipher each block of n message digits as $C = f(m) = AM$, this cipher is linear since $f(M_1 + M_2) = f(M_1) + f(M_2)$. Linear block ciphers are breakable because they are simple substitution ciphers. Now suppose we have a matrix sequence $A^\infty = A_0 A_1 \cdots A_{N-1} A_0 A_1 \cdots$, with period N and $A_i \neq A_j$ for all $i \neq j$ and $0 \leqslant i, j \leqslant N - 1$. Let $X = M_0 M_1 \cdots M_{s-1}$ be s blocks of n message digits. If we encipher X as

$$Y = (A_0 M_0)(A_1 M_1) \cdots (A_{s-1} M_{s-1})$$
$$= f_0(M_0) f_1(M_1) \cdots f_{s-1}(M_{s-1})$$

then this cipher is a polyalphabetic substitution. Although each transform $f_i(M) = A_i M$ is linear, one may expect the cipher to be secure if the period of the pseudorandom matrix sequence is large enough, since within a very long time there are practically no two blocks of message digits which are enciphered with the same simple substitution. But the fact may be not so. For example, the following pseudorandom nonsingular matrix sequence generator consists of n maximum-length LFSRs of length n with the same primitive feedback polynomial $f(x)$ over $GF(2)[x]$.

The initial state vectors of the n maximum-length LFSRs are chosen to be n consecutive state vectors of some maximum-length LFSR with the same feedback polynomial $f(x)$. Recalling basic properties of maximum-length LFSR, we see that each matrix of the output sequence is nonsingular and the output matrix sequence has period $2^n - 1$, which is very large provided that n is large enough, and $A_i \neq A_j$ for $i \neq j$ and $0 \leqslant i, j \leqslant 2^n - 1$. But the sequence satisfies the following linear recursion

$$A_i + c_1 A_{i-1} + \cdots + c_n A_{i-n} = 0 \qquad i \geqslant n$$

where $c_1, \cdots, c_n$ are the feedback coefficients of the n LFSRs. This shows that only a very small number of matrices suffice to linearly determine others. Thus for polyalphabetic substitution ciphers the key stream is not only required to have large period, but also to have small statistical dependency. Since the key stream of a cipher may be a sequence of numbers, matrices and vectors, etc., how can we measure its statistical linear dependency? Noticing that the key streams of many ciphers can be regarded as vector sequences over some field F, we can use the linear complexity of a vector sequence as a measure index on the statistically linear dependency of a key stream.

Let a key stream $k^\infty = k_0 k_1 k_2 \cdots$ be a semi-infinite sequence of vectors of a linear vector space V over a field F, then the smallest positive integer L such that there is a polynomial $f(x)$ in $F[x]$ which satisfies the linear recursion

$$k_i + c_1 k_{i-1} + \cdots + c_L k_{i-L} = 0, \qquad i \geqslant L$$

where $c_1, \cdots, c_L$ are the coefficients of $f(x)$, i. e., $f(x) = 1 + c_1 x + \cdots + c_L x^L$, is called the linear complexity of the key stream. If k^∞ is not periodic, then such an L does not exist; in this case, we define $L(k^\infty) = \infty$. If k^∞ is periodic of N, then $L(k^\infty) \leqslant N$. If $V = F$, then the well known Berlekamp-Massey algorithm provides an efficient way to

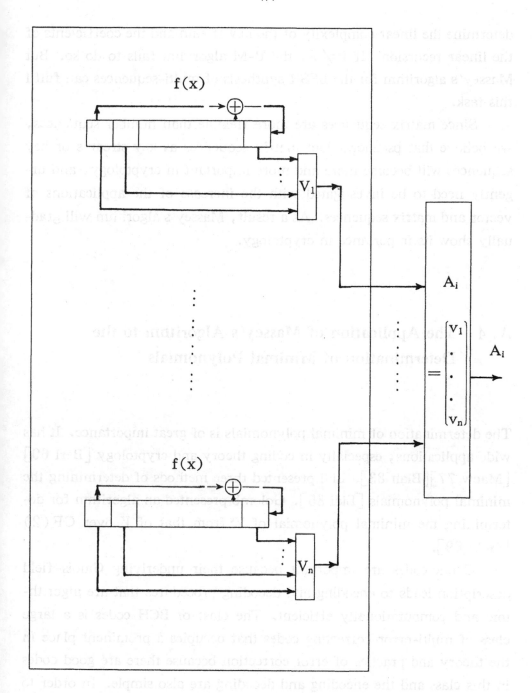

Fig. A. 2 A kind of pseudorandom nonsingular matrix sequence generator

determine the linear complexity of the key stream and the coefficients of the linear recursion. If $V \neq F$, the B-M algorithm fails to do so. But Massey's algorithm for the LFSR synthesis of multi-sequences can fulfil this task.

Since matrix sequences are more flexible than number sequences, we believe that pseudorandom matrix sequences as key streams or key sequences will become more and more important in cryptology, and urgently need to be investigated with the increase of the applications of vector and matrix sequences. As a result, Massey's algorithm will gradually show its importance in cryptology.

A. 4 The Application of Massey's Algorithm to the Determination of Minimal Polynomials

The determination of minimal polynomials is of great importance. It has wide applications, especially in coding theory and cryptology [Berl 69] [Macw 77][Blah 83]. Lidi presented three methods of determining the minimal polynomials [Lidi 85]. Golomb presented an algorithm for determining the minimal polynomial of V^k from that of V over $GF(2)$ [Golo 69].

Cyclic codes are important because their underlying Galois-field description leads to encoding and decoding procedures that are algorithmic and computationally efficient. The class of BCH codes is a large class of multi-error-correcting codes that occupies a prominent place in the theory and practice of error correction because there are good codes in this class and the encoding and decoding are also simple. In order to construct cyclic codes with generating polynomial $g(x)$ which has $\{u_1, \cdots, u_r\}$ as its zeros, we have to find the minimal polynomials of these field elements, denoting them as $f_1(x), \cdots, f_r(x)$ respectively, then calculate $g(x) = lcm(f_1(x), \cdots, f_r(x))$. If we want to construct a t-error-correcting BCH code which has a primitive block length $n = q^m - 1$, the things we have to do are: (1) Choose a primitive polynomial of degree m and construct $GF(q^m)$; (2) Find $f_j(x)$, the minimal polynomial of V^j for $j = 1, \cdots, 2t$; (3) Calculate $g(x)$. Suppose that we have found the minimal polynomials $f_1(x), \cdots, f_{2t}(x)$ in some way, the determination of $g(x)$ from the $f_i(x)$ is still not easy. We now prove that Massey's algorithm can be used to determine the minimal polynomials of field elements and of a set of field elements, and to determine $g(x)$ directly without determining the $f_i(x)$. The algorithm can also be used to determine minimal polynomials of nonsingular matrices over fields.

Let u be a defining element of $GF(q^m)$ over $GF(q)$, then $\{1, u, \cdots, u^{m-1}\}$ is a basis of $GF(q^m)$ over $GF(q)$. Let v and v_i, $1 \leqslant i \leqslant s$, be elements of $GF(q^m)$. Noticing that v^j and v_i can be expressed as

$$v^{i-1} = \sum_{j=1}^{m} b_{ij} u^{j-1}, \quad b_{ij} \in GF(q), \qquad 1 \leqslant j \leqslant m$$

$$1 \leqslant i \leqslant m$$

$$v_k^{i-1} = \sum_{j=1}^{m} b_{ij}^{(k)} u^{j-1}, \quad b_{ij}^{(k)} \in GF(q),$$

we now set

$$B_i = (b_{i1}, b_{i2}, \cdots, b_{im})$$
$$c_i = (b_{i1}^{(1)}, \cdots, b_{im}^{(1)}; \cdots; b_{i1}^{(s)}, \cdots, b_{im}^{(s)})$$

The minimal polynomial of v and that of $\{v_1, \cdots, v_s\}$ can be found by using the following algorithm. The minimal polynomial of $\{v^1, \cdots, v^{2t}\}$ is actually $g(x)$.

In what follows, we shall give a theoretical basis for the above algorithm. The algorithm is based on the following conclusion:

Theorem A. 7 $m(x)$ is the minimal monic polynomial of v over $GF(q)$ if and only if $(m^*(x), deg(m(x)))$ is the shortest LFSR that generates the $B_1 B_2 \cdots B_{m+1}$.

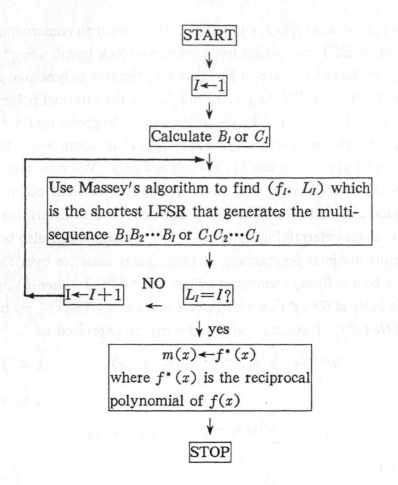

Fig. A. 3   The flow chart of the algorithm for the calculation of minimal polynomial

Proof: Let
$$m(x) = m_0 + m_1 x + \cdots + m_t x^t, \quad m_0 \neq 0, \quad m_t = 1$$
be the minimal monic polynomial of v. By definition we have
$$m_0 v^0 + m_1 v^1 + \cdots + m_t v^t = 0$$
It follows that
$$m_0 v^i + m_1 v^{i+1} + \cdots + m_t v^{i+t} = 0, \quad i = 0, 1, \cdots$$
Therefore, $(m^*(x), t)$ can generate the sequence $B_1 \cdots B_{m+1}$. On the other hand, suppose that $(h(x), L)$ is the shortest LFSR that generates

$B_1 B_2 \cdots B_{m+1}$. By definition we have $h^*(v) = 0$. Thus $m(x)$ divides $h^*(x)$. Thus $m(x) = h(x)$, and $L = deg(m(x))$.

Similarly, we can prove that the algorithm is correct for determining the minimal polynomial of $\{v_1, v_2, \cdots, v_s\}$, i. e. , $g(x)$. Finally, we would like to mention that Massey's algorithm may be used to decode multi-dimension codes.

Appendix B
A Fast Algorithm for Determining the Linear Complexity of Sequences over $GF(P^m)$ with Period P^n

The Berlekamp-Massey algorithm is an efficient one for determining the linear complexity of sequences in general cases. A special algorithm was developed for the determination of linear complexity of a binary sequence with period 2^n by Games and Chan [Game 83]. In the special case, the G-C algorithm works much faster than the B-M. But it does not work for the sequences with period P^n over $GF(P^m)$, where P is prime. The techniques used in deriving the G-C algorithm cannot be easily used to generalize it. In this appendix, we shall use another method to derive a fast algorithm for determining the linear complexity of sequences over $GF(P^m)$ with period P^n. The first part of it is a generalization of the G-C algorithm and the second part is an algorithm for computing the order of root 1 of a polynomial.

Let $N = P^n$, and $s^N = (s_0 s_1 \cdots s_{N-1})$ be the first period of the periodic sequence s^∞. Decompose s^N into P parts of the same length P^{n-1}, say $s^N = L_0 L_1 \cdots L_{P-1}$. The algorithm presented in Fig. B. 1 by its flowchart produces the linear complexity of s^∞ after at most $(P-1)n$ iterations. For example, let $P=3$, $m=1$, $n=3$ and
$$s^N = (100212012221010101111222000)$$
The algorithm works in the following steps:

	s	R	C
step 1)	As given above	102111110	18
step 2)	102111110	021	24
step 3)	021	$s_0 + s_1 + s_2 \neq 0$	27

so $C(s^\infty) = 27$.

The algorithm depicted in Fig. B. 1 is based on the following two theorems.

Theorem B. 1   With L_i as defined above and s^∞ a sequence of period P^n over $GF(P^m)$, let

$$L_i(x) = s_{iP^{n-1}} + s_{iP^{n+1}+1}x + \cdots + s_{(i+1)P^{n-1}-1}x^{P^{n-1}-1}$$

$$i = 0, 1, \cdots, P-1$$

$$R(x) = L_0(x) + \cdots + L_{P-1}(x)$$

and

$$L(x) = L_0(x) + (L_0(x) + L_1(x))x^{P^{n-1}} + \cdots +$$

$$+ (\sum_{i=0}^{P-2} L_i(x))x^{(P-2)P^{n-1}}$$

then set

$$R^\infty(x) = R(x)/(1-x)^{P^{n-1}}$$

and

$$L^\infty(x) = L(x)/(1-x)^{P^n - P^{n-1}}$$

a) If $R(x) = 0$, then $L(s^\infty) = L(L^\infty)$

b) If $R(x) \neq 0$, then $L(s^\infty) = (P-1)P^{n-1} + L(R^\infty)$

Proof: From the definition of $L_i(x)$, $R(x)$ and $L(x)$, we get

$$s^N(x) = L_0(x) + L_1(x)x^{P^{n-1}} + \cdots + L_{P-1}(x)x^{(P-1)P^{n-1}}$$

$$= L(x)(1-x)^{P^{n-1}} + x^{(P-1)P^{n-1}}R(x)$$

Therefore,

$$s^\infty(x) = s^N(x)/(1-x)^{P^n}$$

$$= [L(x)(1-x)^{P^{n-1}} + x^{(p-1)P^{n-1}}R(x)] \times \frac{1}{(1-x)^{P^n}}$$

a) If $R(x) = 0$, then $s^\infty(x) = L(x)/(1-x)^{P^n - P^{n-1}}$. Thus $L(s^\infty) =$

$L(L^\infty)$.

b) If $R(x)\neq 0$, noticing that $deg(R(x))\leqslant P^{n-1}-1<P^{n-1}$, we get from Lemma 5.10 that

$$gcd[(1-x)^{P^s},L(x)(1-x)^{P^{s-1}}+x^{(P-1)P^{s-1}}R(x)]$$
$$= gcd[(1-x)^{P^{s-1}},R(x)]$$

Hence, it follows from Theorem 5.3 that

$$L(s^\infty) = deg[(1-x)^{P^s}/gcd((1-x)^{P^s},s^N(x))]$$
$$= deg[(1-x)^{P^s}/gcd[(1-x)^{P^s},R(x)]]$$
$$= P^n - P^{n-1} + deg[(1-x)^{P^{s-1}}/gcd((1-x)^{P^{s-1}},$$
$$R(x))]$$

$$= (P-1)P^{n-1} + L(R^\infty) \qquad \#$$

Theorem B.2 Let $f(x) = a_0 + a_1x + \cdots + a_tx^t$ be a polynomial over $F[x]$. If $f(1)=0$, then

$$f(x)/(1-x) = b_0 + b_1x + \cdots + b_{t-1}x^{t-1}$$

where $b_i = -(a_{i+1}+\cdots+a_t),0\leqslant i\leqslant t-1$

Proof: Since $f(1)=0$, $(1-x)$ can divide $f(x)$. Suppose $f(x) = (1-x)g(x)$ and

$$g(x) = b_0 + b_1x + \cdots + b_{t-1}x^{t-1}$$

By comparing the coefficients of $f(x)$ and $(1-x)g(x)$, we obtain

$$b_i = -(a_{i+1}+\cdots+a_t), \qquad 0\leqslant i\leqslant t-1 \qquad \#$$

The first part of the algorithm applies the result of Theorem B.1 recursively, and the second part applies the result of Theorem B.2 recursively. Comparing the algorithm in Fig. B.1 with the B-M algorithm, the former works much faster. It only needs at most $(P-1)n$ iterations, but require more storage space. On the other hand, the algorithm presented in this appendix does not need the computation of the inverse of elements on $GF(P^m)$, but the B-M algorithm does. We would like to mention here that the algorithm presented can be modified into a fast algorithm for determining the linear complexity of multi-sequences over $GF(P^m)$ with period p^n.

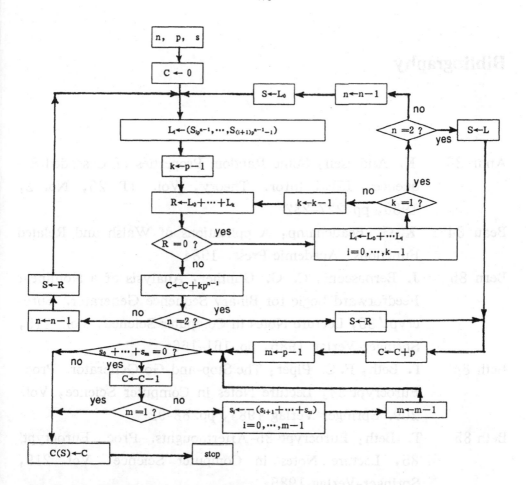

Fig. B. 1 The flowchart of the algorithm for determining the linear complexity of sequences of period p^n over $GF(p^m)$

Bibliography

Andr 80   K. Andresen: Some Random Properties of Cascaded Sequences. IEEE Infor. Theory, Vol. IT- 26, No. 2, 1980, pp. 227-232

Beau 84   K. G. Beauchamp: A pplications of Walsh and Related Functions. Academic Press. 1984

Bern 85   J. Bernasconi, C. G. Günther: Analysis of a Nonlinear Feedforward Logic for Binary Sequence Generator. Eurocrypt' 85, Lecture Notes in Computer Science, Vol. 219, Springer-Verlag 1985, pp. 161-166

Beth 84   T. Beth, F. C. Piper: The Stop-and-Go Generator. Proc. Eurocrypt' 84, Lecture Notes in Computer Science, Vol. 209. Springer-Verlag 1984, pp. 88-92

Beth 85   T. Beth: Eurocrypt' 85-Afterthoughts. Proc. Eurocrypt' 85, Lecture Notes in Computer Science, Vol. 219, Springer-Verlag 1985

Blah 79   R. E. Blahut: Transform Techniques for Error- Control Codes. IBM, J. Res. Devel. Vol. 23, pp. 299- 315, 1979

Blah 83   R. E. Blahut: Theory and Practice of Error Control Codes. Addison-Wesley 1983

Bric 88   E. F. Brickell, A. M. Odlyzko: Cryptanalysis: A Survey of Recent Results. Proc. IEEE, May 1988, pp. 578-593

Brue 84   J. O. Bruer: On Pseudo Random Sequences as Crypto Generators. Proceedings of 1984 International Zurich Seminar on Digital Communications, March 6-8, 1984, pp. 157-161

Cham 84 W. G. Chambers, S. M. Jennings: Linear Equivalence of Certain BRM Shift Regester Sequences. Electronics Letters, Vol. 20, pp. 1018-1019, Nov. 1984

Chan 82 A. H. Chan, R. A. Games, E. L. Key: On the Complexities of de Bruijn Sequences. J. Comb. Theory (A), Vol. 33, 1982, pp. 233-246

Dick 00 Dickson: Linear Groups with an Exposition of Galois Field Theory. Springer 1900

Diff 76 W. Diffe, M. E. Hellman: New Directions in Cryptology. IEEE Trans. on Infor, Theory, Vol. IT-22, Nov. 1976, pp. 644-654

Diff 79 W. Diffe, M. E. Hellman: Privacy and Authentication: An Introduction to Cryptography. Proc. IEEE, Vol. 67, March. 1979, pp. 560-577

Ding 87 C. Ding, G. Xiao, W. Shan: New Measure Indexes on the Security of Stream Ciphers. Proc. Third Chinese National Workshop on Cryptology, Xian, China, 1988, pp. 5-15

Ding 87a C. Ding: Proof of Massey's Conjectured Algorithm. Proc. Eurocrypt' 88. Lecture Notes in Computer Science, Vol. 330, Springer-Verlag 1988, pp. 345-349

Ding 87b C. Ding, G. Xiao, W. Shan: The BAA Attacks on Two Classes of Stream Ciphers. Research Report NWTEI, 1987

Ding 87c C. Ding. W. Shan: On Correlation-Immune Functions. Research Report, NWTEI, 1987

Ding 87d C. Ding, W. Shan: On the Security of Several Kinds of Key Stream Generators. Research Report, NWTEI, 1987

Ding 88a C. Ding: On the Stability of Elementary Symmetric Functions. Research Report, NWTEI, 1988

Ding 88b   C. Ding: Weight Complexity and Lower Bounds on the Weight Complexity of Binary Sequences with Period 2^n. Research Report, NWTEI, 1987

Ding 88c   C. Ding, W. Shan: Lower Bounds on the Weight Complexity of Binary ML- sequences. Research Report, NWTEI, 1988

Ding 88d   C. Ding: Lower Bounds on the Linear Complexity of Non-linear Filtered ML-Sequences Derived from that of Weight Complexity. Research Report, XIDU, 1988

Ding 88e   C. Ding: Lower Bounds on the Weight Complexity of Clock-Controlled Binary Sequences. Research Report, XIDU, 1988, Presented at Auscrypt' 90, University of New South Wales, Canberra, Australia, Jan. 1990

Ding 88f   C. Ding : A Lower Bound on the Linear Complexity of ClockControlled ML- Sequences. Research Report, XIDU, 1988

Ding 88g   C. Ding: Another A pproach to the Stability of Linear Complexity of Sequences. Research Report, XIDU, 1988

Ding 88h   C. Ding: A Fast Algorithm for Determining the Linear Complexity of Sequences over $GF(P^m)$ with Period P^n. Research Report, XIDU, 1988

Ding 88i   C. Ding: Characterizations of Bent Functions. Research Report, XIDU, 1988

Ding 88j   C. Ding: On the Stability of Source Coding for the Sources of Binary Additive Stream Ciphers. Research Report, XIDU, 1988

Ding 89a   C. Ding: Measure Indexes on the Stability of Periods and Their Relationships with Weight Complexity and Sphere Complexity. Research Report, XIDU, 1989

Ding 89b   C. Ding, W. Shan, G. Xiao: The Stability of Period of Binary Sequences with Period 2^n. Research Report, XIDU, 1989

Ding 89C   C. Ding, G. Xiao: Weight Period and the Autocorrelation Function of Binary Sequences. Research Report, XIDU, 1989

Ding 89d   C. Ding: Bounds on the Weight Complexity $WP_k(S^\infty)$ for $1 \leqslant k \leqslant 2$. Research Report, XIDU. 1989

Ding 89e   C. Ding. W. Shan. G. Xiao: The Stability of Period of Binary Sequences with Period 2^n. Research Report, XIDU, 1989

Ding 89f   C. Ding: Ten Open Problems of the Stability of Stream Ciphers. Research Report, XIDU, 1989

Ding 89g   C. Ding, W. Shan: On the Linear-Complexity Stability of Decimated Sequences. Research Report, XIDU, 1989

Diff 88   W. Diffe: The First Ten Years of Public-key Cryptography. Proc. IEEE, Vol. 76, No. 5, May 1988, pp. 560-577

Fen 85   G. Fen and K. K. Tzeng: An Iterative Algorithm for the Multi-Sequences Synthesis with a Shortest LFSR. Sciential Sinica (Science in China), August 1985, pp. 740-749

Game 83   R. A. Games, A. H. Chan: A Fast Algorithm for Determining the Linear Complexity of Binary sequence with Period 2^n. IEEE Trans. Inf. Theory Vol. IT-29, No. 1, pp. 144-146, Jan. 1983

Gare 79   M. R. Garey, D. S Johnson: Computers and Intractability: A Guide to the Theory of NP-Completeness. W. H. Freeman and Co. , San Francisco, Calif. 1979

Goll 84   D. Gollmann: Pseudorandom Properties of Cascaded Connections of Clock Controlled Shift Registers. Proc. Eurocrypt' 84. Lecture Notes in Computer Science, Vol. 209, Springer-Verlag, pp. 93-98

Golo 67   S. W. Golomb: Shift Register Sequences. Holden-Day. San Francisco, Calif. 1967

Golo 69   S. W. Golomb: Irreducible Polynomials, Synchronization Code, Primitive Necklaces, and the Cyclotomic Algebra. Proc. Conf. Combinatorial Math. and its A ppl. pp. 358-370, University of North Carolina Press, Chapel Hill, N. C. 1969

Günt 87   C. G. Günther: A Generator of Pseudorandom Sequences with Clock Controlled Linear Feedback Shift Register. Proc. Eurocrypt' 87, Lecture Notes in Computer Science, Vol. 304. Springer-Verlag 1988

Jenn 80   S. M. Jennings: A Special Class of Binary Sequences", Ph D Thesis, London University, 1980

Karp 76   M. Karpousky: Finite Orthogonal Series in the Design of Digital Devices. J. Wiley, NY and IUP, Jerusalem 1976

Knut 81   D. E. Knuth: The Art of Computer Programming, Vol. 2: Seminumerical Algorithms. Addison-Wesley 1981

Kolm 65   A. N. Kolmogorov: Three A pproaches to the Quantitative Definition of Information. Probl. Inform, Transmission, Vol. 1, 1965, pp. 1-7

Kuma 83   P. V. Kumar, R. A. Scholtz: Bounds on the Linear Span of Bent Sequences. IEEE Trans. Inf. Theory. IT-29. No. 6, 1983, pp. 854-862

Lemp 76   A. Lempel, J. Ziv: On the Complexity of Finite Sequences", IEEE Trans. Inform. Theory. IT-22, Jan. 1976, pp. 75-81

Lidi 85   R. Lidi: Finite Fields. Encyclopedia of Mathematics and A pplications, Vol. 20. 1985

MaFa 73   R. L. McFarland: A Family of Difference Sets in Non-cyclic Groups. J. Comb, Theory, Series A15, pp. 1-10, 1973

Macw 77   F. J. Macwilliams, N. J. A. Sloane: The Theory of Error-Correcting Codes. North-Holland 1977

Mass 69   J. L. Massey: Shift-Register Synthesis and BCH Decoding. IEEEE Trans. on Information Theory, Vol. IT-15, Jan. 1969, pp. 122-127

Mass 85   J. L. Massey: Linear Complexity and Sequence Synthesis. Manuscript of Lecture Notes in Northwest Telecommunications Engineering Institute. Xian. PRC, 1985

Mass 85   J. L. Massey: Coding and Cryptology. Advanced Technology Seminars, Zurich, Aug. 1985, pp. 5. 23-5. 24

Mass 87   J. L. Massey, T. Schaub: Linear Complexity and A pplications. In: G. Cohen, P. Godlewski (eds.): Coding Theory and A pplications. Lecture Notes in Computer Science, Vol. 311, Springer-Verlag 1987, pp. 19-31

Mass 88   J. L. Massey: An Introduction to Contemporary Cryptology. Proc. IEEE, May 1988, pp. 533-549

Mora 79   C. Moraga: Orthogonaltransformationen im Mehrwertigen Logischen Entwurf. In: K. H. Fusol (ed.): Entwurf Digitaler Steuerungen. Springer-Verlag 1979

Mora 85   C. Moraga: On Some A pplications of the Chrestenson Functions in Logic Design and Data Processing. Mathematics in Simulation 27, pp. 431- 439, North- Holland 1985

Olse 82   J. D. Olsen, R. A Scholtz, L. R. Welch, : Bent-Function Sequences. IEEE Trans. IT, Vol. IT- 28, No. 6, November 1982, pp. 858-868

Pich 87   F. R. Picherler: Finite State Machine Modelling of Cryptographic Systems in Loops. Proc. Eurocrypt'87. Lecture Notes in Computer Science, Vol. 304. Springer- Verlag 1988

Roth 76   O. S. Rothaus: On Bent Functions. J. Comb. Theory, Series A20, pp. 330-305, 1976

Ruep 86   R. A. Rue ppel: Analysis and Design of Stream Ciphers. Springer-Verlag, 1986

Ruep 87   R. A. Rue ppel: When Shift Register Clock Themselves.
          Proc. Eurocrypt'87. Lecture Notes in Computer Science,
          Vol. 304. Springer-Verlag 1988

Ruep 88   R. A. Rue ppel. O. J. Staffelbach: Products of Linear
          Recurring Sequences with Maximum Complexity. IEEE
          Trans. Infor. Theory, Vol. IT-33, No. 1, Jan. 1987,
          pp. 124-131

Shan 87   W. Shan: The Structure and the Construction of Correla-
          tion-immune Functions. MS Thesis, NTE Institute, Xian,
          1987

Sieg 84   T. Siegenthaler: Correlation-immunity of Nonlinear Com-
          bining Functions for Cryptographic A pplications. IEEE
          Trans. Inf. Theory, Vol. IT-30, No. 5, Sept. 1984,
          pp. 776-780

Sieg 85   T. Siegenthaler: Decrypting a Class of Stream Ciphers Us-
          ing Ciphertext Only. IEEE Trans. Computers, Vol. C
          034, No. 1, Jan. 1985, pp. 81-85

Sieg 86   T. Siegenthaler: Methoden für den Entwurf Von Stream
          Cipher-Systemen. Diss. ETH Nr. 8185, ADAG Zurich
          1986

Simm 88   G. J. Simmons: A Survey of Information Authentication.
          Proc. IEEE, May 1988, pp. 603-620

Solo 64   R. J. Solomov: A Formal Theory of Inductive Inference.
          Part I. Inform. Control 7, 1964

Tits 62   R. C. Titsworth: Correlation Properties of Cyclic Se-
          quences. Thesis, California Institute of Technology,
          Pasadena, Calif. 1963, pp. 160-170

Tits 64   R. C. Titsworth: Optimal Ranging Codes. IEEE Trans.
          Space Electronics and Telemetry, March 1964 pp. 19-30

Trau 88   J. F. Traub et al. : Information-Based Complexity. Aca-
          demic Press 1988

Voge 84   R. Vogel: On the Linear Complexity of Cascaded Sequences. Proc. Eurocrypt'84, Lecture Notes in Computer Science, Vol. 209. Springer-Verlag 1984. pp. 99-109

Wu 87   C. Wu: Linear Complexity of Sequences. MS Thesis, Northwest Telecommunications Engineering Institute, Xian, PRC, 1987

Wu 88   C. Wu: A Class of New Nonlinear Combining Functions for Stream Ciphers. Third Chinese National Workshop on Cryptology, Xian, PRC, 1988, pp. 54-60

Xiao 85   G. Xiao, J. L. Massey: A Spectral Characterization of Correlation-Immune Functions. IEEE Trans. IT. Vol. 34, May 1988, pp. 569-571

Zeng 87   K. Zeng: The Entropy Leakage in Cryptosystems. The Graduate School of Science and Technology of China, Beijing, PRC, 1987

Lecture Notes in Computer Science

For information about Vols. 1–473
please contact your bookseller or Springer-Verlag

Vol. 515: J. P. Martins. M. Reinfrank (Eds.). Truth Maintenance Systems. Proceedings. 1990. VII. 177 pages. 1991. (Subseries LNAI).

Vol. 516: S. Kaplan, M. Okada (Eds.). Conditional and Typed Rewriting Systems. Proceedings. 1990. IX, 461 pages. 1991.

Vol. 517: K. Nökel. Temporally Distributed Symptoms in Technical Diagnosis. IX. 164 pages. 1991. (Subseries LNAI).

Vol. 518: J. G. Williams. Instantiation Theory. VIII. 133 pages. 1991. (Subseries LNAI).

Vol. 519: F. Dehne. J.-R. Sack, N. Santoro (Eds.). Algorithms and Data Structures. Proceedings. 1991. X. 496 pages. 1991.

Vol. 520: A. Tarlecki (Ed.). Mathematical Foundations of Computer Science 1991. Proceedings. 1991. XI, 435 pages. 1991.

Vol. 521: B. Bouchon-Meunier, R. R. Yager, L. A. Zadek (Eds.). Uncertainty in Knowledge-Bases. Proceedings. 1990. X, 609 pages. 1991.

Vol. 522: J. Hertzberg (Ed.). European Workshop on Planning. Proceedings, 1991. VII. 121 pages. 1991. (Subseries LNAI).

Vol. 523: J. Hughes (Ed.). Functional Programming Languages and Computer Architecture. Proceedings, 1991. VIII, 666 pages. 1991.

Vol. 524: G. Rozenberg (Ed.). Advances in Petri Nets 1991. VIII, 572 pages. 1991.

Vol. 525: O. Günther, H.-J. Schek (Eds.). Advances in Spatial Databases. Proceedings, 1991. XI, 471 pages. 1991.

Vol. 526: T. Ito, A. R. Meyer (Eds.). Theoretical Aspects of Computer Software. Proceedings, 1991. X, 772 pages. 1991.

Vol. 527: J.C.M. Baeten, J. F. Groote (Eds.). CONCUR '91. Proceedings, 1991. VIII, 541 pages. 1991.

Vol. 528: J. Maluszynski, M. Wirsing (Eds.). Programming Language Implementation and Logic Programming. Proceedings, 1991. XI, 433 pages. 1991.

Vol. 529: L. Budach (Ed.). Fundamentals of Computation Theory. Proceedings, 1991. XII, 426 pages. 1991.

Vol. 530: D. H. Pitt, P.-L. Curien, S. Abramsky, A. M. Pitts, A. Poigné, D. E. Rydeheard (Eds.). Category Theory and Computer Science. Proceedings, 1991. VII, 301 pages. 1991.

Vol. 531: E. M. Clarke, R. P. Kurshan (Eds.). Computer-Aided Verification. Proceedings, 1990. XIII, 372 pages. 1991.

Vol. 532: H. Ehrig, H.-J. Kreowski, G. Rozenberg (Eds.). Graph Grammars and Their Application to Computer Science. Proceedings, 1990. X, 703 pages. 1991.

Vol. 533: E. Börger, H. Kleine Büning, M. M. Richter, W. Schönfeld (Eds.). Computer Science Logic. Proceedings, 1990. VIII, 399 pages. 1991.

Vol. 534: H. Ehrig, K. P. Jantke, F. Orejas, H. Reichel (Eds.). Recent Trends in Data Type Specification. Proceedings, 1990. VIII, 379 pages. 1991.

Vol. 535: P. Jorrand, J. Kelemen (Eds.). Fundamentals of Artificial Intelligence Research. Proceedings, 1991. VIII, 255 pages. 1991. (Subseries LNAI).

Vol. 536: J. E. Tomayko, Software Engineering Education. Proceedings, 1991. VIII, 296 pages. 1991.

Vol. 537: A. J. Menezes, S. A. Vanstone (Eds.). Advances in Cryptology – CRYPTO '90. Proceedings. XIII, 644 pages. 1991.

Vol. 538: M. Kojima, N. Megiddo, T. Noma, A. Yoshise, A Unified Approach to Interior Point Algorithms for Linear Complementarity Problems. VIII, 108 pages. 1991.

Vol. 539: H. F. Mattson, T. Mora, T. R. N. Rao (Eds). Applied Algebra, Algebraic Algorithms and Error-Correcting Codes. Proceedings, 1991. XI, 489 pages. 1991.

Vol. 540: A. Prieto (Ed.). Artificial Neural Networks. Proceedings. 1991. XIII. 476 pages. 1991.

Vol. 541: P. Barahona, L. Moniz Pereira, A. Porto (Eds.). EPIA '91. Proceedings. 1991. VIII, 292 pages. 1991. (Subseries LNAI).

Vol. 543: J. Dix, K. P. Jantke, P. H. Schmitt (Eds.). Nonmonotonic and Inductive Logic. Proceedings. 1990. X. 243 pages. 1991. (Subseries LNAI).

Vol. 544: M. Broy, M. Wirsing (Eds.). Methods of Programming. XII, 268 pages. 1991.

Vol. 545: H. Alblas, B. Melichar (Eds.). Attribute Grammars, Applications and Systems. Proceedings, 1991. IX, 513 pages. 1991.

Vol. 547: D. W. Davies (Ed.). Advances in Cryptology – EUROCRYPT '91. Proceedings. 1991. XII, 556 pages. 1991.

Vol. 548: R. Kruse, P. Siegel (Eds.), Symbolic and Quantitative Approaches to Uncertainty. Proceedings. 1991. XI, 362 pages. 1991.

Vol. 550: A. van Lamsweerde, A. Fugetta (Eds.), ESEC '91. Proceedings, 1991. XII, 515 pages. 1991.

Vol. 551: S. Prehn, W. J. Toetenel (Eds.), VDM '91. Formal Software Development Methods. Volume 1. Proceedings, 1991. XIII, 699 pages. 1991.

Vol. 552: S. Prehn, W. J. Toetenel (Eds.), VDM '91. Formal Software Development Methods. Volume 2. Proceedings, 1991. XIV, 430 pages. 1991.

Vol. 553: H. Bieri, H. Noltemeier (Eds.), Computational Geometry - Methods, Algorithms and Applications '91. Proceedings, 1991. VIII. 320 pages. 1991.

Vol. 554: G. Grahne, The Problem of Incomplete Information in Relational Databases. VIII, 156 pages. 1991.

Vol. 555: H. Maurer (Ed.), New Results and New Trends in Computer Science. Proceedings, 1991. VIII, 403 pages. 1991.

Vol. 556: J.-M. Jacquet, Conclog: A Methodological Approach to Concurrent Logic Programming. XII, 781 pages. 1991.

Vol. 557: W. L. Hsu, R. C. T. Lee (Eds.), ISA '91 Algorithms. Proceedings, 1991. X, 396 pages. 1991.

Vol. 558: J. Hooman, Specification and Compositional Verification of Real-Time Systems. VIII, 235 pages. 1991.

Vol. 559: G. Butler, Fundamental Algorithms for Permutation Groups. XII, 238 pages. 1991.

Vol. 560: S. Biswas, K. V. Nori (Eds.), Foundations of Software Technology and Theoretical Computer Science. Proceedings, 1991. X, 420 pages. 1991.

Vol. 561: C. Ding, G. Xiao, W. Shan, The Stability Theory of Stream Ciphers. IX, 187 pages. 1991.

Lecture Notes in Computer Science

This series reports new developments in computer science research and teaching, quickly, informally, and at a high level. The timeliness of a manuscript is more important than its form, which may be unfinished or tentative. The type of material considered for publication includes

– drafts of original papers or monographs,

– technical reports of high quality and broad interest,

– advanced-level lectures,

– reports of meetings, provided they are of exceptional interest and focused on a single topic.

Publication of Lecture Notes is intended as a service to the computer science community in that the publisher Springer-Verlag offers global distribution of documents which would otherwise have a restricted readership. Once published and copyrighted they can be cited in the scientific literature.

Manuscripts

Lecture Notes are printed by photo-offset from the master copy delivered in camera-ready form. Manuscripts should be no less than 100 and preferably no more than 500 pages of text. Authors of monographs receive 50 and editors of proceedings volumes 75 free copies. Authors of contributions to proceedings volumes ar e free to use th e material in other publications upon notification to the publisher. Manuscripts prepared using text processing systems should be printed with a laser or other high-resolution printer onto white paper of reasonable quality. To ensure that the final photo-reduced pages are easily readable, please use one of the following formats:

Font size (points)	Printing area (cm)	(inches)	Final size (%)
10	13.5 x 20.0	5.3 x 7.9	100
12	16.0 x 23.5	6.3 x 9.2	85
14	18.0 x 26.5	7.0 x 10.5	75

On request the publisher will supply a leaflet with more detailed technical instructions or a TEX macro package for the preparation of manuscripts.

Manuscripts should be sent to one of the series editors or directly to:

Springer-Verlag, Computer Science Editorial I, Tiergartenstr. 17, W-6900 Heidelberg 1, FRG

ISBN 3-540-54973-0
ISBN 0-387-54973-0